STUDIES IN IMPERIALISM

general editor John M. MacKenzie

Established in the belief that imperialism as a cultural phenomenon had as significant an effect on the dominant as on the subordinate societies, Studies in Imperialism seeks to develop the new sociocultural approach which has emerged through cross-disciplinary work on popular culture, media studies, art history, the study of education and religion, sports history and children's literature. The cultural emphasis embraces studies of migration and race, while the older political and constitutional, economic and military concerns are never far away. It incorporates comparative work on European and American empire-building, with the chronological focus primarily, though not exclusively, on the nineteenth and twentieth centuries, when these cultural exchanges were most powerfully at work.

Science and society in southern Africa

Science and society in southern Africa

edited by Saul Dubow

MANCHESTER UNIVERSITY PRESS
Manchester and New York

Copyright © Manchester University Press 2000

While copyright in the volume as a whole is vested in Manchester University Press, copyright in individual chapters belongs to their respective authors, and no chapter may be reproduced in whole or in part without the express permission in writing of both author and publisher.

Published by Manchester University Press
Oxford Road, Manchester M13 9NR, UK
and Room 400, 175 Fifth Avenue, New York, NY 10010, USA
www.manchesteruniversitypress.co.uk

Distributed exclusively in the USA by
Palgrave, 175 Fifth Avenue, New York NY 10010, USA

Distributed exclusively in Canada by
UBC Press, University of British Columbia, 2029 West Mall,
Vancouver, BC, Canada V6T 1Z2

British Library Cataloguing-in-Publication Data
A catalogue record for this book is available from the British Library

Library of Congress Cataloging-in-Publication Data
A catalog record for this book is available from the Library of Congress

ISBN 13: 978 0 7190 8048 7

First published in hardback 2000 by Manchester University Press
This paperback edition first published 2009

Printed by Lightning Source

CONTENTS

General editor's introduction — *page* vi
Acknowledgements — viii
Notes on contributors — ix

	Introduction *Saul Dubow*	*page* 1
1	Field sciences in scientific fields: entomology, botany and the early ethnographic monograph in the work of H.-A. Junod *Patrick Harries*	11
2	Making canes credible in colonial Mauritius *William K. Storey*	42
3	A commonwealth of science: the British Association in South Africa, 1905 and 1929 *Saul Dubow*	66
4	'For the public benefit': livestock statistics and expertise in the late nineteenth-century Cape Colony, 1850–1900 *Dawn Nell*	100
5	A mania for measurement: statistics and statecraft in the transition to apartheid *Deborah Posel*	116
6	Police dogs and state rationality in early twentieth-century South Africa *Keith Shear*	143
7	The Race Welfare Society: eugenics and birth control in Johannesburg, 1930–40 *Susanne Klausen*	164
8	Doctors and the state: George Gale and South Africa's experiment in social medicine *Shula Marks*	188
9	Technical development and the human factor: sciences of development in Rhodesia's Native Affairs Department *Jocelyn Alexander*	212

Index — 238

GENERAL EDITOR'S INTRODUCTION

The History of Science has long since emerged from its formerly esoteric status, of interest only to a restricted group of scholars. Three developments have been significant in thus bringing scientific history into the mainstream. The most important has been the realisation that science is as much socially and culturally constructed, and hence politically implicated, as other areas of scholarship. The second has involved definitions of what constitutes science, particularly in its western guise. A supposedly clear distinction between the pure and the applied sciences has become more fuzzy, while areas of endeavour which earlier historians of science might not have recognised as coming within their remit have had to be included, if only because they were regarded as 'scientific' by past generations. The third has involved new relativities. In a post-colonial age, western science can no longer be seen as an all-conquering set of truths, a definer of 'advanced' against more 'primitive' civilisations. While western science is privileged in many ways, through the sheer time, money and effort devoted to it, through its laboratory, experimental and numerical techniques, and through its rapidly developing instrumental technologies, still it is increasingly recognised that it constitutes only one set of ways of looking at natural phenomena. Eurocentric approaches have long since been banished by a recognition of the sophistication of Chinese and other sciences. Yet other modes of scientific understanding, often defined as 'indigenous knowledge', have often humbled the over-confident activities of western imperial science.

The Cape has been a principal point of entry for western science to the African continent from at least the eighteenth century. Astronomy, together with several of the natural and earth sciences, had already been firmly established by the time that Dutch rule gave way to British power. Such activities were carried into the interior by travellers and missionaries intent upon adding to the taxonomic accumulation promoted by western scientific institutions. By the later nineteenth century, such projects had been thoroughly professionalised and the southern African territories constituted an important setting for the operation of various western sciences. After the Boer War, science was seen as one of several regenerating forces for the creation of a newly integrated South Africa under white rule. By this time, museums, scientific societies, educational and learned institutions were well developed to such an extent that the British Association, one form of a 'moving scientific metropolis', found it appropriate to hold its 1905 meetings in the Southern African subcontinent.

But science and medicine were to pursue a surprisingly ambivalent approach to the politics and race relations of the region. There are any number of examples of the sciences being bent to concepts of race hierarchy and manipulated for the maintenance of white political dominance. Yet indigenous

GENERAL EDITOR'S INTRODUCTION

knowledge could also be respected as well as derided. And experiments in social medicine could be world leaders in their field. Such ambiguities are well represented in the essays in this book. Moreover, it is refreshing that this collection is free of the accusations of historiographical isolation which have sometimes been levelled at South African historians. The research represented here not only embraces Southern Rhodesia (Zimbabwe) and Mauritius, but also draws upon parallels elsewhere in the imperial record. The scientific history of the British and other European empires is rapidly growing in sophistication. This book makes a notable contribution to these developments.

John M. MacKenzie

ACKNOWLEDGEMENTS

This volume grew directly out of a conference hosted by the Centre of Southern African Studies, Sussex University, in September 1998. Generous financial assistance from the Centre, as well as from the *Journal of Southern African Studies*, made it possible to bring together experts from three continents for two days of intensive discussion. Rosa Weeks' administrative flair helped to make the event pleasurable and relaxed. Our anonymous external reader made several incisive suggestions for improving the text. Monica Kendall's care and professionalism as copy-editor have been invaluable. I am especially grateful to John MacKenzie for his enthusiastic support, advice and subtle guidance.

Saul Dubow

NOTES ON CONTRIBUTORS

Jocelyn Alexander is lecturer in the department of Historical Studies, Bristol University and a co-author, with JoAnn McGregor and Terence Ranger, of *Violence and Memory: One hundred years in the 'dark forests' of Matabeleland* (2000). She is currently editor of the *Journal of Southern African Studies* and is working on a history of the Zambezi hinterland.

Saul Dubow is reader in history at the School of African and Asian Studies, University of Sussex. In addition to a number of edited volumes, he is author of *Racial Segregation and the Origins of Apartheid in South Africa, 1919–36* (1989) and *Scientific Racism in Modern South Africa* (1995).

Patrick Harries is associate professor of history at the University of Cape Town. Author of *Work, Culture, and Identity: Migrant labourers in Mozambique and South Africa, c.1860–1910* (1994), he has written extensively about the politics of identity and ethnicity in this region, as well as on the work of the missionary anthropologist H. A. Junod.

Susanne Klausen's 1999 doctoral disfertation from Queen's University, Kingston, is about the formation of a national birth-control movement and the establishment of contraceptive clinics in South Africa during the 1930s. She is interested in reproductive politics, past and present, both in southern Africa and in Canada, and has published in the *Journal of Southern African Studies*.

Shula Marks is professor in the history of southern Africa at the School of Oriental and African Studies, London University. She has written about many aspects of South Africa, including health and nursing. Amongst her publications in this field are *Divided Sisterhood: Class race and gender in the South African nursing profession* (1994).

Dawn Nell completed her history honours and masters degrees at the University of Cape Town in 1997–98. She is now a doctoral student at St Antony's College, University of Oxford, working on the history of game farming in South Africa.

Deborah Posel is Director of the Institute for Social and Economic Research at the University of the Witwatersrand, Johannesburg. She has published widely on the nature of the apartheid state and is the author of *The Making of Apartheid 1948–1961* (1991).

Keith Shear is lecturer in African Politics at the Centre of West African Studies, University of Birmingham. His contribution in this volume draws on a recently completed history doctorate from Northwestern University,

NOTES ON CONTRIBUTORS

Chicago, on policing and state formation in early twentieth-century South Africa. He has published in *Gender and History*.

William Kelleher Storey received his doctorate in history from Johns Hopkins University in 1993. Author of *Science and Power in Colonial Mauritius* (1997) and *Writing History: A guide for students* (1999), he is currently a visiting assistant professor of history at Millsaps College, Jackson, Mississippi.

Introduction[1]
Saul Dubow

This volume began with a conference on the topic of Science and Society convened at Sussex University in September 1998. Calls for papers were broadcast on the Internet and in journals in the hope of drawing together scholarly work on a theme of growing general interest. Science was broadly interpreted to include disciplines and fields with claims to scientific method. Although a thorough-going account would have to do so, we did not attempt to define the meaning or boundaries of science or to make categorical distinctions between the 'harder' theoretical and natural sciences and the 'softer' social and applied sciences.[2] In keeping with participants' interests in the social and political context in which science operates, ideological claims made for and on behalf of science were of as much concern as consideration of its actual workings.

So far the history and sociology of science has not been well developed in southern Africa, certainly by comparison with the level of work achieved in India, Australia or Latin America.[3] Most research on the history of science in southern Africa has developed as a by-product of other concerns: with, for example, the political economy of segregation and apartheid, the ideological nature of racial domination, contestations over land, inequitable provision of health services, environmental and agrarian history, and so on. Attention to society, in other words, has preceded interest in the history and philosophy of science. In discussion, conference participants noted this as a constraint, not least because of the absence of detailed internal histories of scientific fields and traditions with which to engage. On the other hand, there was no significant gulf to bridge between historians of science and historians of society. Contemporary scholars of southern Africa have taken for granted the view that science is a socially engaged practice rather than a detached mode of pure or objective research. Moreover, the imprint of Africanist ideas has made it safe to assume that the history of colonial science is rather more than the story of the diffusion of western knowledge into a continental void.

Leaving aside the laudatory accounts of heroic advances in colonial medicine, engineering, the natural sciences, as well as closely focused biographies of individual scientists and their discoveries,[4] most of the critical literature on scientific activity in southern Africa has also tended to presume that scientific knowledge has served as one more powerful tool in the hands of an already powerful colonial or settler ruling elite. This claim is difficult to substantiate in the case of the theoretical and natural sciences, partly because of the dearth of sophisticated historical studies, but also because it is likely that such sciences were not as amenable to political influence as the human and social sciences. In the case of the latter, the view that science operated to serve ruling-class interests is a pervasive assumption. Jonathan D. Jansen's pioneering collection, *Knowledge and Power in South Africa*, is a notable case in point.[5] Elsewhere, studies of the political economy of health and medicine have revealed the role of doctors in maintaining the profits of mining companies to the detriment of workers' health. Research on psychology has shown the role of psychometric and vocational testing in boosting labour productivity, controlling the industrial workforce, and forming assessments of the relative intelligence levels of blacks and whites. Urban planning and housing have been discussed in the context of the segregated division of towns and cities. Histories of social anthropology in South Africa and its Afrikaner nationalist variant, *volkekunde*, have debated the role played by anthropologists in the administration and governance of Africans as well as the elaboration of apartheid theory. Agricultural and botanical science has been considered in the context of rural planning and state intervention in the lives of African peasants. More generally, scientific racism, eugenics and social Darwinism have been shown to have played important roles in justifying white supremacy.[6]

Whether considered in instrumental terms as a direct technique of domination, as a tool of Mammon, or in a more refined Foucauldian manner, as an implicit form of ideological mastery and control, the relationship between scientific knowledge and political or economic power has therefore received considerable attention. This volume offers further evidence of such linkages. Deborah Posel, for instance, argues for the salience of statistics and the 'mania for measurement' in the apartheid state's attempt to manage and control South Africa's various population 'groups'. Also addressing the theme of state rationality and techniques of domination, Keith Shear writes of the specialised use of dogs by police in apprehending black alleged criminals.

INTRODUCTION

And, in the sphere of public health, Susanne Klausen discusses the relationship between the birth control movement and eugenic efforts to improve the racial 'fitness' of whites and to maintain controls over black urbanisation.

The thesis that science is neither neutral nor objective and that it can be utilised in an instrumental fashion to the advantage of those in power is thus not at issue here. But it is not the primary purpose of our collection to elaborate this perspective. Rather, this volume takes its lead from the proposition that science, considered as an ideological discourse, affected rulers as well as ruled. The role of science in sustaining the ideological authority and legitimacy of the already privileged and powerful may indeed have been as significant as its impact upon those formally excluded from structures of power. Claims to scientific knowledge, it is suggested, functioned to enhance the self-image of colonial or settler elites. Moreover, scientific knowledge constituted an attractive ideological repository which competing elements within such groupings, pursuing their own particular or sectional interests, were able to exploit.

From an early stage in the colonial enterprise science was utilised not only to observe, measure and control the 'native other' but to proclaim and shape the self-image of colonisers themselves. Indeed, it may be the case that science played as much of a role in the process of promoting colonial dignity and status in the eyes of the European metropole as it helped in the domination of African subjects. Celebration of the achievements of local scientists, especially those whose work received overseas recognition, was one obvious way in which a sense of national pride and worth could be evoked. But science carried more subtle overtones of worth as well. By virtue of its universalising claims and assumptions science offered a powerful conceptual potential for Europeans to imprint their presence on Africa. It also offered a conceptual means to bridge otherwise bewildering gaps and dissonances between metropole and periphery and, in so doing, to make Africa comprehensible within a European paradigm. In making sense – or nonsense – of exotic 'others', colonial scientists were vitally concerned to validate, affirm and structure their own beliefs and sense of moral or imperial purpose. Comparisons mostly served to glorify European achievements while diminishing Africans, though this was not inevitably or simply the case, as Patrick Harries' chapter on the work of the Swiss naturalist and anthropologist Henri Junod reveals.

Harries shows how European intellectuals saw in Africa images of their own prehistory and societal development. He shows, too, how models and paradigms adopted from the natural sciences could readily be applied to represent and explain African societies in recognisable terms. Evolutionist theories, for example, offered a powerful means of extending understandings of the natural world to discussions about the origins and workings of human social organisation. They offered both a means of making sense of humanity's overall place in the universe and of ranking the position and relative worth of different human groups within that universe. In this regard, the manner in which evolutionist conceptions were applied were to have profound consequences for theories of racial equality or difference.

Also operating within the idiom of comparative progress and evolution, though pitched more directly at the institutional and political domain, Dawn Nell's study of statistical enumeration of livestock and crop production in the nineteenth-century Cape Colony shows how claims to rigorous scientific methods played a role in constructing a positive self-image of colonial development. Reliable statistical information was not only viewed as an essential prerequisite for agricultural improvement and expanded production, it also served to underwrite the Cape colonial administration's view of itself as a force of social and moral progress and to distinguish between progressive Englishmen and 'backward' Boers or Africans. Nell's analysis of the vogue for statistics in the modernising nineteenth-century Cape also offers intriguing parallels – and contrasts – with Posel's discussion of the apartheid state's use of statistics more than half a century later.

The ways in which claims to scientific authority present discursive opportunities for competing interests to secure influence within state institutions is well illustrated in Jocelyn Alexander's discussion of debates within the Rhodesian state around the implementation of the 1951 Land Husbandry Act. Here Alexander shows how a new cadre of technical experts promoted a highly technocratic view of scientific agricultural development which presumed that a centralised state could manage all aspects of agrarian production in the African reserves. This assumption clashed with long-established paternalist traditions within the Native Affairs Department stressing the need to take account of the 'human factor'. Appeals to the authority of an alternative scientific body of knowledge, namely social anthropology, turned the tables as rural resistance and African nationalism increasingly threatened state control. The state declared Africans unready for modernisation and sought instead to court 'traditional authority'; as a result, the hubristic ambitions of the apostles of technical development found their theories displaced by older views sympathetic to the preservation of colonial African social structures, duly recast in terms of the sociology of 'community development' and the need to take account of 'tribal' beliefs.

INTRODUCTION

Alexander's discussion of the place of scientific knowledge in the post-war modernising impulses of the Rhodesian state provides a nice counterpart to Posel's account of the role of statistical science during the heyday of 'grand apartheid' in South Africa. In Posel's view apartheid statecraft placed a tremendous premium on its capacity to count and therefore to control the black populace (a perspective which, incidentally, challenges assumptions that apartheid was a backward-looking movement in which the views of anti-modernising romantic nationalists predominated). State planners sought to utilise their statistical expertise not only to manage processes of labour allocation and urbanisation in the interests of the smooth workings of racial capitalism, but also to confirm to their own supporters that apartheid was a rational system and that the government was the ultimate guarantor of social order.

The relationship between science and state rationality is also a major theme in Keith Shear's chapter on the use of dogs by South African police. In this extraordinary case study, a department of state sought to demonstrate its modernising credentials by developing what it thought of as a sophisticated programme of canine breeding and training designed to apprehend black criminals. The development of an investigative technology based on the supposed abilities of dogs to detect miscreants (and of dog-handlers to manage their charges effectively) was a source of considerable pride. As in the case of Posel's manic measurers, however, such pretensions to knowledge and ordered control revealed an absurd dimension to the policing bureaucracy as leading officials derived comfort and self-assurance by contrasting their own logical practices with the presumed irrationality of African belief systems.

Shear also shows that South Africa's interwar expertise in canine policing was promoted as a source of national prestige. This theme is developed more broadly in my own discussion of the role of science in promoting white South African colonial nationalism. Through comparing the visits to South Africa of the British Association for the Advancement of Science in 1905 and again in 1929, I argue that science was closely associated with the development of white national consciousness and changing notions of imperial belonging. Whereas in 1905 stress was laid on the role that science could play in reaffirming links between South Africa and the British empire and in reiterating the country's status as an offshoot of the colonial metropole, in 1929 emphasis was laid more broadly on the idea of joint participation in the British Commonwealth. This ideological shift reflected changes in the internal configuration of domestic politics as a racially exclusive but ethnically inclusive Anglo-Afrikaner 'South Africanism'

[5]

attempted to command the political centre and to marginalise Afrikaner nationalism. For politicians like Jan Hofmeyr and Jan Smuts, who played a leading role as a Commonwealth statesman, science offered a means to project South Africa as a mature and increasingly self-sufficient member of the international community of nations.

A frequent assumption in critiques of colonial science is that scientific knowledge constituted part of a hegemonic structure of ideological power whose claims to represent progress and enlightenment were primarily a cover for base imperialist motives. It is not necessary to argue the contrary position – that science was value-free and beneficial to all – in order to recognise the limitations of this critique. In the two contributions in this volume concerned with medical science the complexities of this debate are well illustrated. Shula Marks' study of George Gale's advocacy of socialised medicine, during and after the Second World War, cautions against views which assume that all medical knowledge served the interests, directly or indirectly, of ruling elites. That medical science in South Africa has acted in support of the apartheid system is not in question – witness, for example, the role of mine doctors in the management of tuberculosis. But it is difficult to argue this position in the case of social medicine. The championing of community health centres by the likes of Gale and the Karks was motivated by a genuine desire to promote the broad health needs of Africans and their ideas were considerably ahead of their time, not only as far as South Africa was concerned, but internationally too. The failure of the social medicine ideal in South Africa is not surprising given the opposition of the majority of the medical profession, the existence of provincial rivalries and, above all, the social exclusions and fundamental inequalities in South African society. But the story of its limited successes and the fact that progressive South African doctors sought a vent for their ideas overseas (until their ideas were rediscovered by South African health activists in the 1980s) serves as a reminder of South Africa's ambiguous international status during and after the war. It also provides one of many examples in which expertise accumulated at the colonial 'periphery' was exported internationally.

In the case of Susanne Klausen's study of the birth control movement in 1930s' South Africa, the ambiguous nature of medical science is again highlighted. The Race Welfare Society, which forms the focus of her study, was initially founded as part of a eugenic effort to improve the racial qualities of the white race. Its membership comprised both hardline eugenists for whom sterilisation of 'unfit' poor whites was the preferred option, and social reformers who, though accepting of racial segregation in one form or another, viewed birth control as a

INTRODUCTION

benign means of addressing the dire social consequences of urbanisation and impoverishment. For reasons explained by Klausen, it was the latter reforming tendency which gained the upper hand in the Race Welfare Society. It is also of interest that the cause of birth control offered space for women doctors and social activists to assert themselves professionally and to promote the cause of maternal welfare. Class-bound attitudes remained but it is striking that, in spite of the growing forces of racial exclusivism, an organisation which started out in overt support of white supremacy gradually moderated its views towards a more humanitarian position and gradually extended its services to non-white women as well.

The tension between metropolitan and indigenous systems of knowledge has stimulated a large volume of literature in many parts of the colonial world. In particular, the privileging of western science and the displacement of local systems of knowledge by bodies of imperial knowledge has been linked to discussions of the entrenchment of European power, status and authority throughout the imperial domain. This line of argument has helped to stimulate a countervailing appreciation of indigenous knowledge systems, which has in turn been reinforced by the relativist assumptions of postmodernity and a concomitant devaluation of Enlightenment rationality. Debate around such issues has been less developed in South Africa than elsewhere in the ex-colonial world though it is not clear why this should be the case.[7] Indigenous knowledge was undoubtedly effaced or appropriated without due recognition by many – though by no means all – European explorers and scientists. But the evidence suggests that indigenous knowledge *systems* in southern Africa were not able to compete with European scientific traditions as they conspicuously did in countries like India or China. In the absence of sustained research on this topic, this volume does not have a great deal to add on the subject. Nevertheless, a few pointers are worth highlighting by way of stimulating further research and debate.

The presumption of the superiority of European science over indigenous knowledge is apparent in several of the contributions in this volume. Indeed, the exceptions seem to prove the general rule according to which Europeans automatically counterposed their own rationality to the superstitions of indigenous peoples. Shula Marks suggests, for example, that while George Gale retained some openness to the utility of African healers this did not lead him to question the superiority of western medicine, while Patrick Harries points out that although Henri Junod learned much about local systems of biology and entomology from his African informants he did not consider these to be truly scientific. As an anthropologist Junod was sensitive to the existence and worth of different logical schemes and conceptual frameworks but he never departed from the notion that African thought was ultimately governed by magical conceptions (though he did allow for the possibility that they might in time adopt scientific modes of reasoning). Such examples, it should be noted, are drawn from relatively enlightened men of science. More typical, one suspects, were the self-serving attitudes of those who unquestioningly counterposed western rationality with primitive superstition and assumed that such categories were both fixed and racially determined. In the example of Keith Shear's study of sniffer dogs this kind of arrogance meant that the police ironically failed to comprehend that their own pretensions to scientific method were based on serious misconceptions and unfounded beliefs.

Just as there was no homogeneous body of metropolitan knowledge, so there is no sense in which one can talk about indigenous knowledge in unitary terms. It may be useful to distinguish between metropolitan and indigenous knowledge for heuristic purposes but even if such categories are theoretically capable of definition, mutual borrowings, conflicts and interactions frustrate the drawing of rigid distinctions. In the case of established colonial societies, like South Africa or Mauritius, where ruling elites were themselves divided on ethnic and class grounds, attempts to draw simple dichotomies between metropolitan and indigenous knowledge are further complicated (indeed, one of the intriguing aspects of Mauritius is that there are no truly indigenous peoples). Dawn Nell's chapter illustrates how the processes of census enumeration and measurement which were so central an aspect of the nineteenth-century British imperial ideology of progress and enlightenment were often greeted with suspicion by white farmers. Although improving Afrikaner agriculturalists in the western Cape were often sympathetic to the ideology of progress, many others expressed scepticism or hostility towards state attempts to quantify agricultural production. Nell shows that many Boers – like African peasants – were suspicious of expert attempts to interfere with their own tried and tested farming techniques. Over generations, Afrikaner colonisers had become advocates of the superiority of their own vernacular knowledge.

William Storey's discussion of sugar cane production in Mauritius also addresses the intricate debates around metropolitan and colonial science, in this case competing ideas about the propagation of sugar cane advanced by local planters and British and French experts. Science did not of itself disclose the utility of different cane breeds. Instead, political influence, scientific personalities and the demands of the sugar industry all played a role in determining the direction of research

INTRODUCTION

into new cane varieties. The relative status of Franco-Mauritian, Creoles and Indo-Mauritian peasants was also an important factor in gaining knowledge of and access to canes, and so too was the shifting nature of political power. In Mauritius, as in South Africa, local scientific expertise was therefore enmeshed in struggles over political citizenship and national self-assertion.

Finally, a note of caution. Although there was evidently a close relationship between scientific knowledge and colonial power, the nature of this interaction cannot easily be specified or generalised about. For one thing scientific knowledge as well as ideological and political power take different forms. Leaving aside this tricky epistemological problem, it remains difficult to specify precisely how important scientific knowledge was in the ordering of society. Several chapters in this book note the exaggerated claims that were made for the utility of scientific expertise in colonial southern Africa. This may be related to the fact that colonial-based scientists often enjoyed closer access to government than their counterparts in the metropolitan centre. As Richard Drayton has pointed out, 'It was at the Empire's frontier, rather than in London or at Oxford, that the state first became a significant employer of expertise.'[8]

Distance from metropolitan scientific networks and pervasive fears of colonial inferiority or parochialism may also have meant that local scientific and technological successes were over-enthusiastically celebrated as a vindication of colonial progress. But the power and authority associated with scientific expertise was also a reminder of perceived vulnerability and marginality. It was indicative, too, of the limited extent to which colonial officials were in fact able to command authority over subject peoples. Thus, Deborah Posel contrasts the apartheid state's disproportionate ideological investment in statistical knowledge with its conspicuous inability to command full control over processes of African urbanisation; Keith Shear's chapter suggests that the inordinate faith placed in dog tracking reflected the limits of police powers over rural African populations; and Jocelyn Alexander reveals how the failure of technicist agricultural intervention in Rhodesia was conditioned both by the need to concede to chiefly interests and by the continuing ability of Africans to evade and subvert prescriptive demands. A tentative conclusion would be that the allure of science in the colonial societies of southern Africa exceeded its real influence and utility as a technique of domination. But even if its practical and instrumental effects were limited in practice, the valorisation of scientific knowledge for ideological purposes was, and indeed remains, highly salient.

SCIENCE AND SOCIETY IN SOUTHERN AFRICA

Notes

1 In addition to the very helpful comments on this Introduction made by the contributors to this collection, I have benefited from remarks and suggestions made by David Arnold and William Beinart.
2 It seems reasonable to assume that certain sciences were more dependent on metropolitan networks than others and that some were more permeable to local influences and initiatives than others.
3 See, for example, D. Arnold, *Science, Technology and Medicine in India, 1760–1947* (Cambridge, 2000); D. Kumar, *Science and the Raj 1857–1905* (Delhi, 1995); R. Grove et al., *Nature and the Orient: The environmental history of South and Southeast Asia* (Delhi, 1998); T. Griffiths, *Hunters and Collectors: The antiquarian imagination in Australia* (Melbourne and Cambridge, 1996); R. W. Home, *Australian Science in the Making* (Cambridge, 1988); Nancy Stepan, *'The Hour of Eugenics': Race, gender and nation in Latin America* (Ithaca, 1991).
4 See, for example, A. C. Brown, *A History of Scientific Endeavour in South Africa* (Cape Town, 1977). A notable exception is the impressive work of the Cape Town astronomer Brian Warner on the history of South African astronomy and the diverse scientific activities of John Herschel in particular. But the fact that Herschel, one of the country's most important and wide-ranging scientists, has attracted hardly any interest by South African historians, is indicative of the generally undeveloped state of the history of science.
5 J. D. Jansen (ed.), *Knowledge and Power in South Africa: Critical perspectives across the disciplines* (Johannesburg, 1991).
6 For examples of such work, see W. Beinart and P. Coates, *Environment and History* (London, 1995); J. Carruthers, 'Nationhood and national parks: Comparative examples from the post-imperial experience', in T. Griffiths and L. Robin (eds), *Ecology and Empire: Environmental history of settler societies* (Edinburgh, 1997); S. Dubow, *Scientific Racism in Modern South Africa* (Cambridge, 1989); I. Evans, *Bureaucracy and Race: Native administration in South Africa* (Berkeley, 1997); W. D. Hammond-Tooke, *Imperfect Interpreters: South Africa's anthropologists 1920–1990* (Johannesburg, 1997); A. Kuper, *South Africa and the Anthropologist* (London, 1987); R. Packard, *White Plague, Black Labour: Tuberculosis and the political economy of health and disease in South Africa* (Berkeley, 1989); R. B. Millar, 'Science and society in the early career of H. F. Verwoerd', *Journal of Southern African Studies*, 19:4 (1993); B. Nzimande, 'Industrial psychology and the study of black workers in South Africa: A review and critique', *Psychology in Society*, 2 (1984); M. Vaughan, *Curing their Ills* (Cambridge, 1991).
7 The shadow of apartheid is one obvious factor, though its precise impact on indigenous knowledge systems requires careful thought and research. It may be worth noting that the powerful influence of Christianity and of mission education on the African elite in southern Africa led many African intellectuals wholeheartedly to adopt scientific knowledge and 'western' concepts of modernisation as markers of, and guides to, social advancement. Whether such assumptions will now be challenged in the name of an 'African Renaissance' is a fascinating question. For some recent discussions of the interactions between indigenous and colonial beliefs in southern Africa see, for example, J. Peires, *The Dead Will Arise* (London, Johannesburg and Bloomington, 1989); C. Hamilton, *Terrific Majesty: The powers of Shaka Zulu and the limits of historical invention* (Cambridge, Mass., 1998); I. Scoones et al., *Hazards and Opportunities: Farming livelihoods in dryland Africa: Lessons from Zimbabwe* (London, 1996); M. Leach and R. Mearns (eds), *The Lie of the Land: Challenging received wisdom on the African Environment* (Oxford, 1996).
8 R. Drayton, 'Knowledge and Empire', in P. J. Marshall (ed.), *The Oxford History of the British Empire*, vol. II (Oxford, 1998), pp. 249–50.

CHAPTER ONE

Field sciences in scientific fields: entomology, botany and the early ethnographic monograph in the work of H.-A. Junod[1]

Patrick Harries

In 1875 the Encyclopaedia Britannica pronounced Africa to be 'until recently one of the least known ... great divisions of the globe'. Within only a few years, however, readers were faced with waves of knowledge about the continent as European missionaries, traders and scientists moved from their perches on the coast to explore and document the interior. These intrepid travellers returned home with a colourful view of Africa that, by the end of the century, was in its turn being supplemented and superseded by a more ordered or modern gaze.

This new way of seeing influenced the emerging discipline of anthropology. Its exponents in the field, together with their armchair colleagues in the metropole, developed a style of writing that distanced and separated their work from that of earlier genres. Travellers in Africa who aimed their books at a popular readership tended to weave random sightings and indiscriminate observations of local people around an exciting, personalised narrative. Always in search of some Grail, they often portrayed Africans as obstacles along the route or as anonymous 'pagans' whose beliefs and customs impeded the spread of Christianity. In contrast, early anthropologists attempted to explain these customs and beliefs and, in the process, situated indigenous people at the centre of their concerns. At the same time they attempted to make their writing more objective and scientifically authoritative by withdrawing themselves from the text.[2] This strategy of allowing the facts to 'speak for themselves' was supported by the widespread use of terms and concepts drawn from both classical scholarship and indigenous knowledge, and by the methodology, language and conceptual imagery of the natural sciences.[3]

In this chapter I examine ways in which the methodology of entomology and botany influenced the beginnings of anthropology in southern Africa. By tracing the early life history of one particularly influential scientist turned anthropologist, the Swiss missionary Henri-Alexandre Junod, I hope to achieve two objectives. Firstly I want to determine how the form, content and authority of the early ethnographic monograph was shaped and contained by the conventions of writing and analysis of the natural sciences. I particularly want to examine how the skills of observation developed by field naturalists in Switzerland were transferred to the new discipline of anthropology. I am also concerned with theories imported from Europe, and developed locally, with which early anthropologists made sense of their data. Although Junod claimed to be a simple man-on-the-spot, his ethnography was built on innovative notions of fieldwork and theory that were deeply rooted in the experiences of natural scientists in Switzerland and Germany. In South Africa, Junod amalgamated the teachings of his European mentors with the explanations of his English-speaking colleagues. And in the process he created the theory and methodology on which anthropology was built in the region – some thirty years before the 'revolution' brought to the discipline by imperial anthropologists.[4]

Neuchâtel and natural science

Henri-Alexandre Junod was born into a family blessed by the knowledge and practice of science and religion. His father Henri was born in the canton of Neuchâtel in 1825, the last of eleven children of a farm labourer. When Henri senior left his small village for the town of Neuchâtel, the closed pews of the twelfth-century collégiale church were still reserved for the elite; and the establishment of a bourgeois Republic in 1848, which brought independence from Prussia and full incorporation into the Swiss Confederation, brought little change to this strict social hierarchy. Henri combined his interest in theology with a passion for the study of nature. This was largely because he saw nature as the product of God's hand but also, perhaps, because of a remarkable group of naturalists who, during the 1830s–1840s, turned the small cantonal capital into an international centre of scientific research.

The polymath Frédéric de Rougemont was one of the earliest to return from the Prussian capital where he had studied under the geographer Karl Ritter. But most famous was Louis Agassiz who, with the support of Georges Cuvier and Alexander von Humboldt, moved home to teach at Neuchâtel in 1832. He soon built up an internationally renowned research team. Amongst these Arnold Guyot, who had studied under Ritter and von Humboldt in Berlin, would go on to a long and distinguished career at Princeton (1855–83), Carl Vogt would hold the rectorship of the University of Geneva and Edouard Dessor, who stayed in Neuchâtel to become professor of geology at the Academy, would play a leading role in the intellectual and political life of the canton. Agassiz would move to Harvard to become 'the most powerful and imperious biologist in America,' and an inspiration to later generations of aspirant natural scientists in Neuchâtel and elsewhere.[5]

Part of the renown of these extraordinary scholars emerged from their willingness to spend extended periods of time in the Alps where they recovered fossil remains, or measured the structure and movement of glaciers and observed the transformation of névés into ice. Their findings pushed back the age of the world and filled it with a startling diversity. But their studies left humanity dangling on the edge of a monstrous gap bordered on the one side by the act of Creation and on the other by biblical and classical scholarship. To fill this void, Dessor later turned to archaeology and de Rougement to the investigation of primitive society. During the 1850s interest in this work increased palpably as engineers cut the first railways through the mountains of Switzerland. As they drilled through different geological strata, a series of fossils was discovered that allowed a closer reading of the prehistoric past. At the same time, a severe drought lowered the water level of some of the major lakes in the country and uncovered extensive lacustrine villages dating back to the Stone Age. Henri Junod's generation obsessively gathered evidence on the origins and development of the physical world and the place occupied by humanity in its long history. This deep concern with the relationship between geology, society and history was transmitted to his son's generation, many of whom, as missionary explorers trained in Greek and Latin, would see in Africa a reflection of Europe's prehistory and, more distantly, the primitive origins of humanity.

In 1861 Henri Junod was posted to Chezard-Saint-Martin where his first son, Henri-Alexandre, was born two years later. As the village pastor, Junod taught natural history, particularly geology and ornithology, in the local school. One of the young people who fell under the spell of this charismatic clergyman was a young primary school teacher named Fritz Tripet. Self-taught and from a humble background, Tripet would in 1873 become, with Junod and others, a founder member of the Independent Church of Neuchâtel; and a decade later, equally under the influence of Junod, he would become the professor of botany at the town Academy.[6]

As a man of religion and science, Henri Junod moved into the upper echelons of cantonal society. Able and articulate, he had married into a family of leading local industrialists, was elected in 1867 to the prestigious manse of the collégiale church in Neuchâtel, and secured the directorship of the town's Sunday School movement. He also took care to provide his children with the social and cultural capital that had served him so well.[7] Junod introduced young Henri-Alexandre to the botanical diversity of the collégiale's garden and encouraged him to collect and classify plants. With the help of his sister Elizabeth, Henri-Alexandre collected plants in the Jura and pasted them into his first formal register.[8] His love for natural history blossomed when he entered Neuchâtel's classical *collège latin* which housed the town's outstanding natural history museum and counted among its staff an eminent, Berlin-trained teacher and researcher in the field of zoology, Paul Godet.[9]

As a young student in Neuchâtel, Henri-Alexandre Junod benefited from the patronage of Fritz Tripet who, as the secretary of the Natural History Society founded by Agassiz, the editor of its journal (1879–1907) and soon the professor of botany at the Academy, became a locally influential figure. He encouraged students to search out and bring him plants which he named; and in general, he emphasised what we might today call 'fieldwork' over and above the more abstract questions raised by plant anatomy.[10] He was also able to oversee the publication of Henri-Alexandre's first articles in zoology and botany.[11]

At college Junod was strongly influenced by George Godet, a professor of theology who introduced him to Kant, and particularly by George's brother Frédéric, another of Ritter's students, who served as professor of exegesis and criticism. In an intellectual climate shaken by the recent discoveries of Darwin and others, these pious Calvinists and their *émigré* friends in America and elsewhere battled against conservatives who saw science merely as a breeding ground for the scepticism and rationalism undermining biblical revelation. From America, Agassiz saw 'manifested in the animal and vegetable kingdoms... the thoughts of the Creator of the Universe'.[12] Arnold Guyot and his close friend Frédéric Godet wrote influential works in which, following the tradition of George Buffon, they reinterpreted the seven days of biblical Creation as a natural process extending over seven stages or epochs.[13] They believed human reason, as much as personal revelation, demonstrated the existence of God. For Godet science was 'a sister, a powerful ally of faith'.[14] 'The light of religious revelation and science shine from different origins,' he wrote in the manner of Francis Bacon, 'one comes from the sky and the other from the earth, but in meeting they combine to produce perfect clarity'.[15] For Guyot 'the book of Nature' was as much the work of God as the Bible; both were a source of divine knowledge and, 'coming from the same Author, complement one another, forming together the whole revelation of God to man'. From this perspective, the study of nature revealed 'innumerable proofs of the almighty power and wisdom of its author'.[16] The glimpse of the Celestial Pilot provided by nature was even clearer in the pulpit, from where Henri-Alexandre's father thundered against 'the obscurantists, the stiflers, the gagging priesthood,' who questioned the material and social benefits brought by science. Like many of his friends, Henri believed that a thorough knowledge of science would reinforce the girders of Christianity and draw to it 'the noblest of thirsting souls'.[17]

These Christians found their religious beliefs confirmed as they came to view the act of Creation as less the product of a single divine *fiat* somewhere in a remote past and more as an ongoing process linking the past and present, and extending into the future. In this way God was not, as St Augustine had first speculated, a transcendental figure beyond the stars who had created the world and then left it to its own devices; He was rather an all-pervading presence whose immanence and omniscience could be read in the series of creations that constituted the world in both time and space. This perspective caused earnestly severe men like Henri Junod and Frédéric Godet to view nature with a sensibility and spirituality that could move them to tears when reading a weighty scientific tome.[18] As the study of nature emphasised the patent glory of God, many eminent evangelicals in French-speaking Switzerland, particularly those associated with the missionary movement, devoted themselves to its study.[19]

For these men the study of nature was a means of discovery, a source of virtue, a mark of good taste and an emotional release from an otherwise rigorous lifestyle. Their view of nature was not, however, accepted uncritically in all quarters of the town. Many held steadfastly to the biblical narrative of Creation and, in viewing an organism solely as the product of a divine designer, ignored its relationship to the surrounding environment. In Neuchâtel this perspective produced the sort of natural history undertaken by Samuel Robert (1853–1934) who saw lepidoptera purely as an expression and proof of God's glory. Through purchase and exchange, Robert assembled a magnificent collection of 23,000 butterfly species from all over the world. But while Robert was only interested in the remarkable beauty and diversity of these insects, a collector such as Frédéric de Rougemont's son Frédéric (a pastor in the Independent Church) sought to discern the laws behind this diversity. To achieve this, he hunted and raised butterflies, noted their habits and sought, in the narrow confines of one specific region, to uncover a classificatory 'system' based on the Linnaean model.[20] He also sought to infuse the youth with his enthusiasm and when his brother, the professor of natural history at the Academy, died unexpectedly, de Rougemont and Tripet arranged for Henri-Alexandre to continue his research into microlepidoptera. In 1884 the twenty-one-year-old Junod published his findings on the anatomy, habitat and customs of a tiny butterfly in the scientific journal founded by Louis Agassiz some forty years earlier.[21]

At college, Henri-Alexandre became head student of the faculty of theology where he simultaneously led the local chapter of the intellectually and socially important Société de Belles-Lettres. In the society's *Revue* he published poetry and articles on regional history and literature; but he was particularly renowned for natural history essays that, through a combination of sentimentality, information and moral philosophy, saw an uplifting spirituality in God's handiwork. His readings could hold an auditorium of *belletriens* spellbound. 'Junod,' wrote an admirer,

> whose gaze knows how to penetrate the mysteries of nature and discover movement and life where the greatest scientists have seen only inertia and immobility, Junod who recognizes a soul under the icy exterior of a drop of frozen rain and who believes he is able to perceive in the soul of a small tree frog or a seagull religious and moral instincts and a yearning for the ideal.[22]

Towards the end of his degree he spent a semester at the University of Basel before going on to the University of Berlin. When Junod arrived in the German capital a young scientist, equally influenced by Kant, natural history and the disciples of Karl Ritter, had only recently returned from Baffin Island. This was Franz Boas who had finished his *habilitation* and was preparing to embark on a career in North America just as Junod completed his own degree and returned home to enter the clergy in October 1885. Both men would take a particularly Germanic intellectual cocktail into the field of Anglo-American anthropology.

When Junod was posted as a missionary to Lourenço Marques in 1889, he carried with him a love of the natural and theological sciences typical of the dominant male figures in his life. He saw the post to which he had been sent by his vocation as not only an opportunity to convert the heathen and revivify his church at home; it would also allow him to discover new botanical and zoological species.[23] It was a good time for an energetic young man to seek his fortune overseas for there was little left to discover in Switzerland where collectors had been reduced to discovering subspecies or at best filling in the gaps. Perhaps Junod was aware that, as early as the middle of the century, an ambitious botanist such as Pierre-Edmond Boissier (1810–85) had been obliged to look to Spain and later the Middle East and the edges of India to discover new plants. Boissier's willingness to move to the edges of Europe and beyond had paid handsomely. In a manner that set a shining example for younger botanists to follow he had described, alone or with the aid of collaborators, a phenomenal 131 new genera of plants and 5,990 new species. His son-in-law William Barbey (1842–1914) continued this tradition through the establishment of the Boissier Herbarium in Geneva which pursued a policy of purchasing botanical collections, sponsoring expeditions and publishing the findings of men-on-the-spot.[24] Entomologists like Auguste Forel were also obliged to look beyond Switzerland in their search for new species.[25] Perhaps most immediately for Henri-Alexandre, just as he discovered his microlepidoptera did not represent a new species, his sister Ruth sent the first of a series of rare natural history specimens from the Transvaal to Godet at the Museum in Neuchâtel.[26]

The natural sciences were also experiencing a rapid transformation at the end of the nineteenth century. For decades the *homme de cabinet* or armchair expert and the amateur collector in the field had existed in an easy symbiosis. In Neuchâtel the growth of the Natural History Museum had traditionally depended on the generosity of the town's many traders, missionaries, mercenaries and travellers living abroad. In a way that reinforced their ties with home, these men sent 'curiosities' to their compatriots in the museum. But the gap between the collector and the classifier widened into something of a generational gap when, following the spectacular achievements of Darwin and the rise of the 'new biology' in the German universities, botanical and zoological excursions lost ground to theory. In the last third of the century, innovators in the field shifted from collecting and documenting the majestic diversity in nature to revealing the causes and consequences of that diversity. Technological developments also encouraged the move from the field to the laboratory as increasingly powerful microscopes allowed professionals to pierce below the surface and specialise in morphology, embryology and physiology. In some cases this led them to classify and recatalogue plants and insects according to minutiae previously not visible to the human eye. Meanwhile, the sheer volume of species discovered rendered the old, encyclopaedic museum collections impracticable and forced experts to specialise in narrow fields. As prestige and status passed to the scientist, the thrill of discovery shifted away from the painstaking but plodding, 'antiquarian' work of collectors like Tripet and de Rougement.[27] Even an internationally renowned botanist such as Boissier came under fire for his unwillingness to ask more general and abstract questions of his material.[28]

Religion underlay much of this shift in thinking. If conservatives had seen the work of Agassiz and his colleagues as leading towards an unsavoury deism, the direction taken by science at the end of the nineteenth century pointed towards an even less savoury atheism. The social and economic instability accompanying industrialisation was compounded as biologists and geologists reformulated species and with it the clear distinctions between the sexes, the animal and plant kingdoms, and even the existence of humanity as a separate species. For many Darwinians, evolutionism publicised the arbitrary nature of classification as it challenged the idea that living objects' anatomical characteristics were fixed and constant through time. As science overturned established ideas on the fixity of species and the place of humanity in the universe, it inevitably cast doubt on the story of Creation and even the existence of God.

As a believer by temperament and an amateur with a broad appreciation of the natural sciences, Junod was better suited to be in the field in Africa than in the laboratory in Europe. Leaving Southampton on 16 June 1889, his excitement rose noticeably as he neared Africa. When the *Tartar* stopped at Lisbon he was captivated by the exotic plants he found in the city's botanical gardens.[29] Three weeks later the ship docked at Durban for two days where he was charmed by the 'luxuriant winter vegetation'. It 'would be impossible', he exclaimed, 'to describe the pleasure that we experienced in the presence of this new nature: flowers, butterflies, all unknown until that moment!'[30] Three months after his arrival in Mozambique Junod had to reassure the mission council in Lausanne that the passion with which he pursued natural history in his spare time had not dented his evangelical zeal. The biological and theological sciences were organically tied, he reminded his superiors, for 'nature is the work of God and merits attention and calls for study'.[31]

Collecting in Africa

For many years only the most intrepid naturalists had visited Lourenço Marques, a backward corner of a forgotten colony.[32] But when Junod arrived, the town was entering a period of unprecedented prosperity. As the natural port for the newly discovered Witwatersrand gold fields it drew to it a diverse and cosmopolitan population. This included people like Rose Monteiro who earned an income by selling zoological and botanical specimens to collectors in Europe.[33] But Junod felt that he had been transported into 'an entirely new world', 'a completely new natural environment, full of surprises and still virgin'.[34]

FIELD SCIENCES IN SCIENTIFIC FIELDS

Three months after arriving at the Rikatla mission station, Junod wrote to William Barbey, the director of the Boissier Herbarium. Although he specialised in the Near and Middle East, Barbey had purchased several important herbaria from southern Africa.[35] He was keen to expand his institute's holdings in this area and viewed missionary societies as a means of achieving this end.[36]

Junod offered to send Barbey both dried plants and the seeds of species suitable for cultivation in hot houses or gardens where they would 'flourish . . . beneath the sky of the fatherland'.[37] Ambitious and energetic, he took just sixteen months to gather over 300 different plant species. Starting from the hill above Lourenço Marques, he explored the 23-kilometre stretch of land between the town and Rikatla in the north. From his mission station he spent his spare time investigating the groves of palm trees and marshy hollows lying behind the coastal dunes where he found a new plant soon to be named the *Striga junodii* Schinz. Evangelical itineration allowed him to scour the local countryside and took him to many different botanical areas. These included the Makororo forest on the route from Rikatla to Antioka where he found the *Empogona junodii* Schinz, the nearby Manununu forest covering the bend in the Nkomati valley, and the rocky Libombo mountains on the border with the Transvaal.[38] An hour to the north of Rikatla, the Morakwene forest covered the south bank of the Nkomati estuary from the coast to the drift as well as two small islands in the river. With its dense underbrush and thick creepers, he found this impenetrable forest an 'eldorado' for the naturalist. Junod was particularly attracted by a tree orchid whose striking burgundy coloured flowers appeared after the first summer rains; and he quickly gave his name to this plant, Annonaceae *Monodora junodii*. He was also impressed by the Makandja forest on the left bank of the Nkomati estuary as its dark clay soil supported a rich and varied flora.[39]

One of Junod's first steps on his arrival at his mission station was to establish what he called, with some irony, 'a museum'. In this centre he dried plants which, in January 1891, he despatched for the first time to Geneva where Barbey undertook to have them named, classified and described in an article tentatively entitled 'Plantae Junodiae'.[40] The twenty-eight-year-old missionary was 'very proud' when told that his first consignment contained at least one unknown species which would thenceforth bear the name *Melhania junodii* Schinz.[41] As further encouragement, Barbey supplied Junod with the equipment needed to establish a herbarium, as well as with information on how to dry plants, particularly during the subtropical wet season, and how to ensure their safe delivery to Europe.[42] He also offered to act as an agent for the sale to museums and collectors in Europe of all Junod's plants not required by his herbarium. With Barbey's encouragement Junod assembled collections holding 100 different species of lichen, fruit, flowers and roots. With the money earned from the sale of these plants he trained and employed an assistant to dry and press specimens and hired the services of locals to search out rare species, particularly those found in malarial swamps or at some distance from Rikatla.[43] When Junod left Mozambique for Switzerland in 1896 he had inscribed 500 different plant species in his botanical register.

Three years later he returned to Africa to take up a post at the school for evangelists at Shiluvane in the eastern Transvaal. The flora of the Delagoa Bay area paled in comparison with that of Junod's new home.[44] Shiluvane was at the heart of a rich botanical area encompassing the Drakensberg and adjoining plateau, the thick forests on the eastern slope of the mountain, the foothills in which the mission was situated, and the hot Lowveld leading down to the coastal plain. Shiluvane occupied a central position in the Swiss mission field and on trips to Pietersburg, via the Woodbush and Haenertsberg, and to the Spelonken and Komatipoort, Junod was able to enlarge and diversify his botanical collection. At the height of the malarial summer Junod and his family spent several weeks every year in the highlands at Howick in Natal or at Thababosigo in Lesotho. Here he found the time to devote himself to his botanical collection – and to look with envy at the unexplored flora of the Maluti mountains. When he returned to Rikatla from Shiluvane in 1906 he had added over 2,300 plant species to the original 500 in his collection.[45]

During the ten years Junod had spent in Switzerland and the Transvaal, the Delagoa Bay region had been thoroughly botanised. The Portuguese had sent an official to collect plants and Rudolf Schlechter, a renowned collector and future professor of botany in Berlin, had spent four months scouring the area.[46] Sadly for botany but happily for anthropology, this competition encouraged Junod to specialise in the study of human beings and, over the next decade, he inscribed fewer than 100 plants in his register. But once he had established his reputation in the field of anthropology Junod returned with relish to botany as a pastime. In the three years before he returned definitively to Switzerland in 1920, he collected a further 1,550 different plant species drawn from many parts of southern Mozambique as well as Sibasa, the Spelonken, Johannesburg, Witzieshoek and Pinetown in Natal.[47]

Henri-Alexandre Junod discovered many new plants during his years in southern Africa and is today recognised as one of the pioneer botanical explorers of the region. Apart from Barbey, he corresponded with various botanists to whom he sent material. These included John Briquet, the director of the Conservatoire Delessert in Geneva, a botanical institute housing numerous South African collections;[48] Fritz Tripet and Paul Godet in Neuchâtel; J. M. Wood, the director of the botanical gardens in Durban; Joseph Burtt Davy, the government botanist in Pretoria; and Thomas Durand, the director of botanical gardens in Brussels.[49] The plants he gathered in the Delagoa Bay region were described in 1899 in two articles co-authored with Hans Schinz.[50] Schinz described his other plants in a series of papers, published between 1893 and 1931, that placed Junod amongst the greatest collectors of African flora of his generation.[51] His memory was perpetuated by the genus *Junodia* Paxis of the family Euphorbiaceae (later sunk into synonymy with *Epinetrum delagoense*) and some thirty different plant species. His plant specimens are today housed in herbaria throughout the world.[52]

Despite this success in the field of botany, Junod found the Delagoa Bay region a richer field for entomology and it was in this area that he initially invested most of his energy.[53] Although Lourenço Marques lay on the edge of the tropics, the hill above the town was arid and the meagre vegetation produced few flowers. But on this bush, he wrote with pleasure in 1891, 'live an astonishing and surprising collection of fauna,' particularly caterpillars.[54] At Rikatla he found it more difficult to trace rare species; but within two years he had trained 'intelligent natives' to recognise exotic insects and uncover their habitat. In the Morakwene forest he found rare caterpillars and, in spring and summer, large numbers of butterflies of the genus *Papilio*, followed by *Charaxes* in autumn and winter. In a field of cashew trees hosting rare bushes and tropical vines lived a world of swallowtail and dawn butterflies, as well as various other endemic species and subspecies.

Junod sometimes raised butterflies and moths from their form as caterpillars; but more frequently he resorted to employing assistants to chase these insects from their habitat by beating the bush at specific times of day or night. His assiduousness brought quick results. In November 1890 he discovered a butterfly (*Teracolus calais*) in the Lower Nkomati valley that previously had only been associated with tropical areas. In the same year he found five examples of the magnificent *Precis cuama* at Howick in Natal and the first example of this species in the Delagoa Bay area. A few months later, one of his assistants brought him a magnificent unclassified swallowtail butterfly – soon to take the name *Papilio junodi*. It took Junod just seven years to assemble some 200 different species of moths and 184 species of butterfly. When placed alongside the 16 other species found by Rose

Monteiro, Junod's collection of butterflies from south-east Africa amounted to almost half the total thought to exist in the entire sub-continent at that time.[55]

From his 'museum', Junod assiduously sought out and studied bugs, crickets, grasshoppers, ants, wasps, bees, moths, molluscs, lizards, frogs and especially caterpillars, beetles and butterflies. He sold rare butterflies to the museum in Neuchâtel and to collectors such as Roland Trimen, the celebrated lepidopterist and director of the South African Museum in Cape Town.[56] In an attempt to identify and name his specimens he corresponded with experts in Lausanne, Geneva, Zurich, Berlin, Bucharest, Turin, Paris, Caen, London and Cape Town. Lepidoptera (butterflies and moths) were despatched to Frédéric de Rougement, Trimen and A. G. Butler at the British Museum. In Lourenço Marques he met and later corresponded with the distinguished entomologist W. L. Distant. Coleoptera (beetles) went to Dr Bugnion in Lausanne and especially Louis Péringuey at the museum in Cape Town; Orthoptera (crickets, grasshoppers and cockroaches) and Hymenoptera (wasps, bees and ants) to Dr von Schulthess-Rechberg in Zurich (where they seem to have been classified by Forel); Hemiptera (bugs and lice) were sent to Professor Montandou in Bucharest; and caterpillars to Mr Heylaerts, a Dutch specialist.[57]

Henri-Alexandre Junod's name is inscribed in the halls of southern African entomology. During his first furlough in Switzerland (1896–99) he worked through his entomological collections and co-authored a series of articles with European experts. In a work on the beetles of the Delagoa Bay region, written with E. Bugnion of Lausanne, he listed 479 species, including eight new species in the family Tenebrionidae, and described many of their localities and habits. This led Bugnion to compare the entomological wealth of Delagoa Bay with the vaunted fauna of Brazil and India.[58] In other articles published in 1899 Junod noted hundreds of insects in the Orthoptera, Hemiptera and Hymenoptera families, several dozen of which were new discoveries.[59] His collection of Hydrophilidae was analysed and described in articles by Dr Griffini of Turin and Dr Regimbart of Evreux.[60] Louis Péringuey was also full of praise for Junod's work on Coleoptera, especially those found around Rikatla, and in 1904 published an article on his findings at Shiluvane.[61] Junod hoped that his many scientific articles would one day be assembled in a single volume, tentatively entitled *The Natural History of Delagoa* or *The Flora and Entomological Fauna of Delagoa*.[62] But his interest in natural science was overtaken by his new passion for anthropology. Nevertheless, he left an indelible mark on the entomological map of south-east Africa where scores of insects bear his name, that of his wife, or the many stations of the Swiss mission.[63]

Throughout his stay in southern Africa Junod supplied the Natural History Museum in Neuchâtel with a wide range of animals and insects. The museum sent him flasks, dissecting cases, cyanide and nets; and it frequently paid for the items collected or for their delivery. This policy paid handsome dividends. The museum register records receiving from Junod in 1911 'a precious collection' of Hemiptera (the bugs and lice for which he had a special fondness), and 187 Orthoptera and Hymenoptera the following year. His molluscs were also considered 'extremely interesting' and were accompanied by a steady stream of sea shells, bird skins, eggs and nests, sea urchins, snakes and lizards, several frogs, a crocodile and various (unidentified) mammals.[64] By appealing to men on the spot like Junod, as well as other Neuchâtelois resident overseas, the museum built up a mass of objects that could later be sifted through by the experts.[65] Junod concurred with this ethos. 'In collecting all these facts,' he remarked, 'we accomplish a sort of highly dignified ministry and, perhaps, we furnish those who will systematize them one day with the key to many problems.'[66]

Explaining collections

Just as biology and theology overlapped, botany and zoology frequently imbricated when, for instance, Junod stumbled upon a little known bush housing several new species of Longicorns; or, in drying a plant for his herbarium, he found an unusual caterpillar lodged on its underleaves. Biology and sociology also frequently overlapped when Junod, like his mentors at home, turned to native informants for information on local plants or insects. This led him to develop a deep respect for their powers of observation and an admiration for their system of classification. He also had a high regard for the way in which local people exploited plants for qualities that were at once medicinal, magical, sartorial and, in edible or liquid form, nutritional. He was intrigued by the way trees were associated with deities and spirits, and plants and animals were employed in rituals, as talismans or as a means of divination.[67]

Elias Libombo or 'Spoon' proved a particularly adept butterfly collector. Because of his ability to find rare butterflies, and distinguish between them, Spoon was frequently sent on expeditions to isolated areas.[68] Junod was also impressed by the way children gave names to beetles when they admired their shape, colour and taste. He was equally pleased to find a 'notion of *order*' in the natives' classification of mammals.[69] Although the natives did not group animals into large classes such as genera, they usually classified animals into smaller categories, if only because some were considered taboo or physically disgusting.[70] They also speculated on man's place in the physical world when they saw apes as a degenerate form of humankind.[71] Junod was particularly impressed by the ability of the locals to classify plants into the equivalent of broad genera and even species. The botanical knowledge of the natives was the product of 'a true and, in a certain sense, scientific observation on their part'. It was certainly 'more general' than that of the European peasantry, perhaps comparable to 'that of our forefathers of two or three hundred years ago, before Botany became a true science'. Even children, he observed, could point out roots used for medicinal purposes.[72] This was fortunate, for Junod was coming under increasing pressure to demonstrate the utility, explanatory value and didactic benefits of his entomological and botanical studies.[73]

Junod found that the indigenous people had a great deal to teach European scientists about the utility of plants. On several occasions he assembled groups of natives in his museum where, in exchange for a payment of 1 shilling, they supplied him with the local names and uses of plants. He particularly admired the botanical knowledge of old women and *nangas* or specialist medicine men. The *nanga*, he came to recognise, were familiar with 'real, powerful drugs' which they administered in conjunction with therapeutic practices.[74] Because Junod regarded botany as a contemplative pastime subordinate to his evangelical duties, he refused to engage in speculations on whether coffee could be established as a cash crop in the Delagoa Bay area. But he did attempt, with little success it seems, to exploit the pith of a local palm as a commercial substitute for cork, and he made genuine attempts to see if medicinal cures could be coaxed from local plants.[75]

Although Junod looked increasingly to local knowledge to explain the natural environment, he still saw 'God's infinite wisdom' behind the diversity of nature. 'Nature is a book, in Africa as in Europe', he wrote in 1893, and a means of praising 'the beauty and power of the Creator'.[76] Four years later he still believed, at least publicly, that natural history provided 'the occasion to admire the magnificent work of the Creator on the marvellous planet that we inhabit'.[77] But even as he reiterated this belief, Junod was further disengaging the Creator's hand from the order of nature.

By the end of the century Junod stressed, to scientists at least, that as a collector he was no longer just concerned with anatomy and classification. His major interest had become the place of plants and animals in the human and natural environment. This shift in interest had been occasioned by his growing engagement with the theory of evolution through natural selection.[78] In Cape Town Roland Trimen was at the centre of a circle of collectors who sought, in the process of natural selection, an explanation for the customs and habits of animals and plants; and he quickly introduced Junod to the works of evolutionists like A. R. Wallace and H. W. Bates. Another important influence at this time was Lord James Bryce, who met Junod at Lourenço Marques and become his principal patron in the world of anthropology.[79] Bryce portrayed Africa as both a menagerie for fauna extinct elsewhere in the world and as an endlessly sprawling, unkempt botanical garden. He attributed this abundance of nature to 'the fact that the country was occupied only by savages, who did little or nothing to extinguish any species nature had planted'. And he went on to speculate that this 'may have caused many weak species to survive when equally weak ones were perishing in Asia at the hands of more advanced races of humankind'.[80]

W. L. Distant felt that Africa was a privileged site from which to view the process of natural selection. He believed that, because the vegetation of south-east Africa had been little disturbed by human agency, it harboured species of butterfly that had been destroyed by more industrious races elsewhere in the world. Hence more species of butterfly were found in southern Africa than in the entire European continent; and Britain's sixty-six species were the meagre vestiges of a far richer age. Distant believed that lepidoptery would provide, like palaeontology, a window into prehistoric times. 'These pleasant Durban glades,' he wrote, 'were no longer only emporiums to supply museum drawers with specimens, but were full of nature's record of the past – like hieroglyphic writings, but unlike them, most at present we cannot read.' A cabinet of butterflies, he asserted, 'now not only exhibits what used simply to be called the "works of nature," but absolutely in many cases shows how nature works'.[81]

Under the influence of these world-renowned entomologists Junod soon started to see natural selection at work in nature, particularly in the field of entomology. In order to more clearly distinguish and classify butterflies, he focused on the reproductive cycle of the insects, particularly the various stages of their metamorphosis. He quickly noted the relationship between insects and a changing, rather than pristine, environment. In many areas the butterfly *Papilio corinneus* had been decimated because the bush on which it laid its eggs was an important source of native medicines. But the same butterfly proliferated in the Morakwene forest where the underbrush proved impenetrable to human agency. He noted that only the fittest butterflies and moths survived as they were preyed upon incessantly by spiders, birds, rodents, mantises, lizards and frogs; and their caterpillars were parasitised by wasps and flies or consumed by human beings.[82] To preserve themselves in such a way as to reproduce their species, Junod realised that insects had evolved various protective mechanisms. This was clearly visible when a specific species of caterpillar curled itself in such a way as to resemble the excrement of a bird, or another hid itself in a sheath of sand or leaves.

Most startling was the mimicry practised by butterflies and moths. Prodded by Trimen and others, Junod came to see that only those egg-laying females survived whose coloration and design conformed most closely to that of butterflies considered malodorous, unpalatable or even toxic by predators. Through this natural selection the insects seemed to practise an involuntary form of deceptive mimicry. Other butterflies developed a protective camouflage when, over generations, only those survived whose design and colour conformed to elements of the local environment. Through a close comparison of the spectrum of variation in a species, Junod was able to document the degree to which these insects adapted physically, albeit involuntarily, to new floral and faunal kingdoms. He noticed that, when a species of butterfly left its home area and moved to an area to which it was not adapted, the butterflies declined in number, strength and size, or were made extinct. He saw how in one area a female butterfly would 'mimic' the colour of a local, poisonous butterfly in such a way as to protect itself against predatory birds. Yet in a neighbouring area, where the birds did not exist, members of this species of butterfly would keep their original colouring and design as they had no need to practise this 'mimicry'. The same process was observable when the clearing of vegetation caused a poisonous or malodorous butterfly to emigrate or die and, in this manner, render the mimicking species unprotected. Similarly, when a change in climate caused the dark lichen on the pale bark of a tree to die, dark butterflies lost their 'natural' camouflage and became an easy prey to predators.

Through these entomological observations Junod gauged the effect of rapid changes in the physical world upon the involuntary mutation or evolution of what he started to call 'Darwinian species'. The process of 'natural selection,' he came to recognise, was the product of a 'pitiless' struggle in the animal world.[83] But while he could accept Darwinian ideas about the evolution of plants and animals, Junod was, perhaps understandably, only able to apply these ideas to humankind in a selective manner.

For Mary Barber, one of Trimen's circle, natural selection had fundamentally challenged the notion of divine providence, the Christian tenets of humility, love and compassion, and the special place of humanity in the order of the world.[84] For Darwin, natural selection was a purely mechanical phenomenon that, devoid of all purpose, was marked by pitiless indifference to the human condition. Although human beings could influence the process of evolution through their ability to think and choose, the effects of natural selection were entirely random and gratuitous. Junod found it difficult to apply these laws to the notion of human development. For if eternal salvation was the aim of life on earth, the process of evolution had to be subordinate to individual morality and divine purpose. Junod's entomological studies brought him to see evolution as the product of natural selection rather than the successive special creations envisaged by Agassiz and others.[85] But in his adoption of Darwinian ideas, Junod attempted to retain the basis of nineteenth-century Christian humanism. He would carry this combination of natural science, philosophy and theology into his understanding of anthropology.

Science and anthropology

When Junod moved his attention from entomology and botany to anthropology, he carried the impersonal, authoritative style of the scientific genre into his ethnographic writing. In the preface to his monographs he was careful to draw attention to his linguistic abilities, the skills of his informants and the reasons for producing what he described as 'a collection of biological phenomena which must be described objectively'.[86] But his disengaged, scientific style inevitably came into conflict with his deep concern for the welfare of the local population. 'An African tribe is not just an object of study like the birds, animals or insects displayed in the windows of our museums and dissected by diligent scientists,' he wrote, straining against the conventions of the scientific genre. 'A tribe is a living thing.'[87] To resolve the conflict between the humane missionary and the detached scientist, the compassionate romantic and the objective scholar, he placed his subjective commentaries in passages 'carefully separated from the scientific treatise'. It was 'out of respect for science' that he relegated topics such as 'alcoholism and the South African tribe' or 'the place of the vernacular in native education' to the appendices of his ethnographic monograph.[88] Junod also wrote extensively about social change in his private correspondence, mission reports and essays; and treated the topic in various works of fiction, including short plays and a full-length novel. But in his published, scientific work he excluded any reference to the traumatic effects on indigenous society of capitalism and colonialism.

In the United States Agassiz had come to define humankind as a genus composed of several fixed reproductive units or species, each made up of a variety of races. He thought Africans were, like animals, polygenic in origin and incapable of reproducing themselves through intermarriage or miscegenation with other human 'species'.[89] Junod's adoption of the notion of evolution through natural selection reinforced his religious belief that all human beings were members of a common reproductive unit and that the origins of humanity could be traced to a monogenetic source.[90] But his observation of the effects of evolution on variation within animal species led him to look for the same process in human development. He came to see that culture was not only related to the environment, as Ritter had postulated; it was also the product of different levels of reason and thought that were reflected through language. This led him to divide humanity into 'civilisations' and 'races' that were coterminous with broad Indo-European, Bantu and other linguistic categories.

Junod's training in classical philology and Bible history, as well as his predilection for Romanticism, told him that a primitive race such as the Bantu should be divided into tribes. He and his missionary colleagues easily determined the linguistic content and borders of these tribes. But the creative process involved in the construction and assemblage of a standard written language was obscured when they claimed to have 'discovered' languages in the same way as naturalists had 'discovered' species of beetles and plants. Because Junod saw languages evolving over time in the same way as species, he viewed the suppression of oral dialects by a written standard as a process of natural simplification that would recreate the linguistic cohesion of the tribe. This would in turn reconstitute a shared cognitive unity that, he believed, would be further strengthened by collections of folksongs and tales that reflected the primitive or 'pure' mentality of the tribe. Linguistic research particularly 'resembled that of the palaeontologist,' he ventured, for 'man's history, his migrations, is revealed to he who investigates the languages and traditions of primitive people'.[91]

In portraying the tribe as an organic unit, Junod supplied it with a history, physical presence and soul, a life of its own and a capacity to reproduce itself. It was around this concept of the tribe as a unit capable of reproduction, that Junod constructed chapters on 'the evolution of the man,' 'the life-cycle of the woman' and even the narrative structure of the first volume of his *Life of a South African Tribe*. The second volume began with a description of the natural environment and its relationship to what Adolf Bastian in Berlin called the 'psychic' or mental life of the tribe.[92] The tribe was certainly the main character in an ethnographic plot that tended to dissolve other actors into roles and types that, like distinct organs, each with its appropriate function, made up the overall anatomy of the tribe. Photographs of 'native salt manufacture' or 'consulting the bones,' contributed to this picture by representing individual behaviour as general practice. The same effect was achieved when individual creations, such as huts, carvings or weapons, were transformed into standard 'Thonga' types. This image of the tribe was reproduced and reinforced when photographs of individuals were inscribed with the prefix 'Thonga' despite the fact that the people depicted were probably unaware of the word and were almost certainly ignorant of its externally imposed meaning as an indicator of identity. The dominance of the tribe was such that the voices of individual informants, so stridently presented in the preface to his monographs, were ultimately submerged by a phalanx of faceless 'Thongas', all anonymous experts on tribal culture.

This view of the tribe as a 'natural' entity created a relativistic approach to culture. The stress on cultural difference was reinforced by Junod when he advised that artefacts be arranged in museum displays according to tribal origin rather than evolutionary typology. He particularly favoured the adoption of dioramas that had proved so successful in exhibiting biological specimens in their natural habitat. 'An ethnographical sample without the knowledge of its origin,' he wrote, 'is no more use than a fossil without the indication of the locality or the geological stratum from which it comes.' By embedding artefacts in their cultural settings, Junod hoped to provide museum visitors with indigenous meanings; but he equally valued tribal arrangements as a means of comparing levels of cultural evolution. The power of museum displays, he felt, could produce 'at a short glance, more comprehension of the Kafir life than many descriptions in ethnographical books'.[93] Photographs, too, could be used to capture verisimilitude; and as 'a faithful, unprejudiced witness,' he used them liberally to illustrate his monographs.[94] Cartography also helped create a vision of the tribe as a 'natural' community; for once the tribe was inscribed on paper, it was broken into various dialect zones to which, again fusing language and primitive politics, Junod gave the kinship term, 'clans'. These 'clans' were presented on the map as clearly defined tribal segments highlighted by sharp lines and primary colours. For Junod, the culture of the tribe inhabiting these severely delimited spaces was as homogeneous as the plant or insect species that filled the biologists' distribution maps.

By fixing the tribe in time and space in this manner, Junod could engage in detailed fieldwork in a limited area and legitimately present his findings as representative of the tribe as a whole. His vision of the tribe as a biological organism also encouraged him to subject it to the insights of direct observation, the methodological convention that he had used to describe zoological and botanical species. But when applied to the study of humans, this concern with only the moment of discovery produced an ethnographic present that attached little importance to the dating of evidence.

Junod's experience in the domains of entomology and botany quickly led him to apply the skills of the field naturalist to the study of humanity. Methodologically ethnographic work could 'be done only by the co-operation of two different agencies,' he wrote, 'those who are to collect the materials, and those who are to work them out'. In the manner of the entomologist in the field, the 'man on the spot' was to observe the object of study in its natural surrounding; and his job was simply 'to note the facts carefully and to describe them accurately'.[95] The professional anthropologist at home was also in the position of his homologue in the natural sciences, for his task was to compare, classify and systematise findings from all over the world, and devise theories to explain their existence.[96] As a man on the spot whose vocation precluded long spells away from the verandah, Junod turned for information to a series of native assistants. These ranged from incidental scholars and herdboys to paid informants, foremost amongst whom was his trusted butterfly collector Elias 'Spoon' Libombo.[97]

Junod's remark about the roles separating the fieldworker from the metropolitan expert should not be taken at face value, for he was never a simple collector of data.[98] He was anxious that his work should supply magistrates, missionaries and others with the native logic and indigenous understanding underlying local customs and beliefs; and it was for this reason that he turned for explanations to locally knowledgeable people.[99] As a former diviner, Elias Libombo did not simply supply Junod with information on his erstwhile occupation; he also explained how divining operated and why people believed implicitly in its powers. Junod's respect for local knowledge increased as he deepened his familiarity with the language, extended his fieldwork and grasped the different forms of logic that underlay indigenous systems of botany or divining. He stressed the need 'to enter into the mind of these primitive men,' and he tried to capture their ideas through a liberal use of vernacular terms in his ethnography.[100]

Junod analysed his data with the tools of diffusionism, and the rudimentary cultural relativism first developed by Ritter, Bastian and other German scholars. In the United States these ideas would triumph under the aegis of Franz Boas. In South Africa the stress on culture as the prime marker of difference would contribute to the *volkekunde* developed at the Afrikaans universities; but in English-speaking circles this way of thinking initially had less success. This was partly because of the tenacity of the evolutionist hypotheses held by the university professionals who dominated anthropology in the British empire until the 1920s. But it was also because Junod sought to make sense of his ethnographic observations by turning to the Darwinian theory of natural selection that had explained so much of his entomological work.

For Junod, the object of anthropology and palaeontology imbricated with that of lepidoptery when he wrote that Europeans could learn about their lost prehistory through the study of primitive peoples; or when he compared the pre-Roman tribes of Europe with those of modern Africa.[101] By drawing analogies between anthropology and palaeontology he both reinforced and created the idea that Europe's prehistory could be recovered from the African present. This view was confirmed by his representation of the vegetation and animal life of south-east Africa as 'entirely new,' 'still virgin,' and full of 'mysteries' waiting to be 'unveiled'. In this supposedly 'pristine' natural environment he had discovered a new world of plants and animals; and he believed he could, in the same way, discover a 'pure' population marked by simple virtues and uncorrupted values. Hence Africa was an exemplary location in which to apply the methodology of the field naturalist to the study of humankind in its most primitive state. But to see in Africa's present a reflection of Europe's lost innocence was to have an enormous impact on his conceptualisation of the discipline of ethnography. It both endorsed his view of anthropology as a salvage operation and confirmed his evolutionist ideas.[102] This led him to believe that Africa's isolation from the centres of world Enlightenment had both retarded the continent's development and protected it from the vices of industrial civilisation. But at the end of the nineteenth century these vices threatened to engulf Africans who, unlike butterflies, had not been hardened by an extended struggle against invasive predators. The black man 'has not been moulded by a civilization where the struggle for survival is pursued without truce or respite,' wrote Junod.[103] This was a defenceless world into which a corrupt European civilisation had suddenly brought slavery, alcoholism and greed. Venereal disease, TB, prostitution and homosexuality were spreading from the cities, and money was eating at the ligaments of tribal social control. Without the ability to defend itself, primitive society was, like certain entomological and botanical species, increasingly threatened by a retrograde movement in the process of evolution toward simplification, degeneration and extinction.[104]

This was particularly visible to Junod when natives moved from their rural homes to industrial cities. As an entomologist interested in the geographical distribution of forms, Junod had supplied de Rougement with butterflies for comparison with members of the same species found in Switzerland. Just as the small size of the African variants indicated an incomplete adaptation to the local environment, so too did the movement of tribesmen to the towns. In this insalubrious, impersonal environment primitive people were exposed to all the vices of European civilisation and, like the humble butterfly, could deteriorate in strength or even suffer extinction. But where Junod, like Lamarck, believed that a weak or inferior society could change, adapt and fortify itself through rivalry and competition, others read weakness and inferiority as inherited, inflexible and biologically determined.[105] Junod's field observations had led him to believe that in some areas, such as linguistics and oral literature, the culture of the natives was superior to that of Europeans; and as we have seen, he found the natives' general knowledge of botany and zoology often more thorough than that of European peasants and workers.[106] Although he criticised phrenological studies claiming to prove Africans' biological inferiority, Junod remained a rather disinterested agnostic in the field of physical anthropology.[107] In some quarters this led to criticism of his attempt to group people in 'artificial' tribes rather than classify them scientifically according to physical type.[108]

Junod's ranking of society in class terms sometimes slipped into a simple racial hierarchy in which Africans occupied an inferior position. Here again, science played a distinctive role for, in the final instance, the natives exhibited 'the want of an enlightened botanical sense'. This was partly because their botanical knowledge was overly utilitarian. If related species were not considered materially useful, they were given only the most generic of titles. Thus, while Junod was able to distinguish between over twenty types of lichen, the locals merely called these *bulele*; and in a similar way, they referred to all ferns generically as *tsonna*.[109] Alternatively, if plants were named, this was often for unscientific reasons as they were considered useful in attracting good fortune or warding off evil.[110] Perhaps most importantly, the natives based their naming of plants on external characteristics alone. This allowed them to recognise broad similarities and differences between plants but restricted their ability to distinguish between the numerous species within a genus. They had never dissected a flower, and were ignorant of the existence of male and female organs. Because they had no idea of the anatomy of flora, the notion of genus or the relationship between species was not always 'correctly and universally applied'.[111]

The zoological knowledge of the indigenous people was, Junod believed, still at a 'very primitive stage'. It was perhaps equal to that of Leviticus, the figure associated with taboos in the Old Testament.[112] As most insects were not eaten, they were called by only the most general names; all butterflies bore the title *phaphalati* and all the black beetles living in sandy areas were named *shifoufounounou*. Although women of the Nkuna clan distinguished between animals with hoofs and those with paws, they did this merely because of a superstition according to which the latter were wild beasts whose meat was fit only for men. While it was rare to find a native who could trace a species of moth or butterfly to a particular type of chrysalis, insect pests or those associated with local 'superstitions' were named, as were almost all (edible) birds.[113]

The natives' knowledge of botany and zoology showed they possessed powers of reasoning and were capable of improvement. But this reasoning was dominated by superstition, magic and despotic customs patrolled by autocratic chiefs. 'The fundamental difference between the European and the Bantu mind,' wrote Junod, was that Europeans have the *'scientific spirit* and the Bantus *the magic conception of Nature.'* For Junod, the notion of cause and effect constituted almost a litmus test of civilisation. Here again, he combined evolution and culture to create primary markers of social difference. But these differences were not racially determined, for the natives would sooner or later recognise the inferiority of their 'strange, unscientific ideas' and adopt the 'scientific knowledge' that was conquering the world.[114] 'Let the great modern principle of experimental science be instilled into their minds,' proclaimed Junod, 'and all that scaffolding of superstitions, which appear to them most reasonable now, will tumble down at once.'[115] Although scientific knowledge initially entrenched racial domination, it quickly became a resource available to all those able to acquire its teachings and willing to adopt its principles. A man like Raúl Honwana thought his teacher Henri-Alexandre Junod 'a great man of learning' with an impressive 'science laboratory in his house'. And like his teacher, Honwana built the fortunes of his powerful family on the forms of knowledge associated with science and religion.[116]

Conclusion

In this chapter I have attempted to show how Henri-Alexandre Junod employed the methods of the natural sciences to represent and explain African society. Entomology and botany provided him with a fine eye for detail, a series of hypotheses with which to make sense of his material, and the narrative conventions needed to order his ethnographic data in an authoritative way. His experience as a field naturalist encouraged him to attract informants to the verandah and to observe life at first hand in the villages. It also brought him into contact with a sophisticated local knowledge of plants and animals that fuelled his respect for cultural difference and fired his curiosity to know more about the lives of his informants. Junod's education in Berlin and Germanophile Neuchâtel provided him with notions of cultural relativism and diffusionism that he fused, in an original way, with the Darwinian evolutionism of South African entomologists and the social evolutionism of Anglo-American anthropology. As a French-speaking foreigner in southern Africa, and as an amateur without an institutional base or intellectual clients, Junod could not stray far from the writings of Morgan, Tylor, Frazer, Haddon and Marrett. The fact that French reviewers of his first monograph, like Emile Durkheim and Marcel Mauss, criticised his lack of theoretical rigour and his salvage fieldwork, merely pushed him closer to the British anthropologists of empire.[117] Under their patronage, Junod's ethnography became a standard work of reference in the new anthropology departments at South African universities.[118]

Politicians were also impressed by the infallible authority with which Junod depicted a primitive society threatened by the materialism of an industrial age. But while some saw in his scientific treatise the evident need for protective segregation, some found in it a clear set of grievances stemming from the destruction by capitalism and colonialism of the cultural authenticity of a golden age.[119] Perhaps ironically, just as Junod's salvage anthropology provided segregation with an intellectual underpinning, his scientific work came under fire from a new generation of professional anthropologists. These university-based intellectuals were concerned to turn anthropology into 'a natural science of human society' by ridding it of the conjectural and speculative history associated with notions of cause and effect.[120] The revolution brought to the discipline by Radcliffe-Brown and Malinowski, and the stark political implications of representing Africans as primitive tribespeople, pushed the work of Junod's generation into the shadows of the discipline.

From our vantage point in time we may perhaps look back on his work with a more detached eye. The demise of legalised segregation and apartheid has drawn the sting from the political implications of his work. The extent of the epistemological break claimed by Radcliffe-Brown today seems less clear. Building on the work of Junod's generation, Radcliffe-Brown believed that social structures could be classed and compared according to their outward appearance, in the same way as sea shells. This led Edmund Leach to compare him to a 'butterfly collector' who, without any sense of his role in the construction of classificatory typologies, chose at whim to distinguish and arrange 'butterflies according to their colour, their size or the shape of their wings'.[121] It is not just the classificatory system of Radcliffe-Brown and his followers that has come under fire; for functionalism is today regarded as merely one more generalising theory alongside the social evolutionism of Junod and his contemporaries. The extended fieldwork in small-scale societies advocated by Malinowski is also seen as an overly narrow methodological base. In today's open intellectual climate, scholars will return with a fresh approach to the writings of Junod's generation, writings that employed the authority of science to build the foundations of a modern image of African society.[122]

Notes

1 This chapter owes a great deal to Pat Lorber of the Bolus Herbarium at the University of Cape Town and Christian Dufour of the Natural History Museum is Neuchâtel. I also extend my thanks to Patrica Davison, Betty Louw and Harmish Robertson of the South African Museum, Cape Town.
2 J. Vansina, 'The ethnographical account as a genre in Central Africa', *Paideuma*, 33 (1987); W. MacGaffey, 'Ethnography and the closing of the frontier in Lower Congo, 1885-1921', *Africa*, 56:3 (1986); R. Thornton, 'Narrative ethnography in Africa, 1850-1920: The creation and capture of an appropriate domain for anthropology', *Man*, 18:3 (1983).
3 For work on this theme, see H. Kuklick 'Island in the Pacific: Darwinian biogeography and British anthropology', *American Ethnologist*, 23:3 (1996); J. Urry, 'From zoology to ethnology: A. C. Haddon's conversion to anthropology', in his *Before Social Anthropology: Essays on the history of British anthropology* (Chur, Switzerland, 1993).
4 W. D. Hammond-Tooke's *Imperfect Interpreters: South Africa's anthropologists 1920-1990* (Johannesburg, 1997) omits the beginnings of the discipline and confirms the chronological division found in A. Kuper, *Anthropologists and Anthropology: The modern British School* (London, 1983).
5 S. J. Gould, *Bully for Brontosaurus* (London, 1991), p. 312. The botanist Léo Lesquereux also moved to an important career in the United States.
6 M. de Tribolet, 'Fritz Tripet: professeur de botanique à l'académie: 1843-1907', *Bulletin de la société neuchâteloise des sciences naturelles* (henceforth *BSNSN*), 35 (1909), 92, 99; Henri Junod [*père*], *Sermons* (Neuchâtel, 1884), p. ix.
7 H.-A. Junod, 'Fritz Courvoisier et sa famille en 1831', *Musée neuchâtelois* (1912), 89-121; University of South Africa (UNISA), Junod Collection (JC), 1.1 biographical notes on H.-A. Junod.
8 Botanical Conservatory, Geneva; Boissier Herbarium (henceforth CBG.HB), Junod to W. Barbey, 2 January 1892.
9 Swiss Mission Archive (henceforth SMA), Lausanne: 503 Junod to secretary, 30 October 1889. On Paul Godet, see G. Dubois, *Naturalistes neuchâtelois de XXe siècle* (Neuchâtel, 1976), pp. 49–51.
10 De Tribolet, 'Fritz Tripet', p. 94.
11 'Le Triton lobé', *Le Rameau de Sapin*, 13 (August 1879), 31–2, 35–6; 'L'Erythronium dens canis.Linné', *Le Rameau de Sapin*, 16 (November 1882), 41–2.

12 L. Agassiz, *Contributions to the Natural History of the United States of America* (1859) vol. I, p. 135, cited in E. Mayr, 'Agassiz, Darwin and Evolution', in his *Evolution and the Diversity of Life* (Cambridge, Mass., 1976), p. 256.
13 F. Godet, 'Les Six jours de la création', in his *Etudes bibliques* (Neuchâtel and Paris, 1889); A. Guyot, *Creation or the Biblical Cosmogony in the Light of Modern Science* (New York, 1884).
14 V. Rossel, *Histoire littéraire de la Suisse romande* (Neuchâtel, 1903), p. 606.
15 Godet 'Les Six jours', p. 88
16 Guyot, *Creation*, pp. 3–4, 7.
17 'Les obscurantistes, les éteignoirs, la prêtraille', in H. Junod, *Du manque de pasteurs et des moyens d'y remédier* (Neuchâtel, 1864), pp. 17, 23–4, 29, 40. Junod *Sermons*, p. 186.
18 Junod, *Manque de pasteurs*, p. ix. See also P. Godet, *Frédéric Godet: 1812–1900* (Neuchâtel, 1913), pp. 82, 334.
19 These included the professor of geology at the Lausanne Academy, Eugène Renevier, who served on the major governing bodies of the Mission for almost forty years, and as its president from 1883 to 1906; the naturalist and former missionary in India Auguste Glardon (a member of the Mission council 1869–73 and 1881–89); and the famous botanist William Barbey (council member from 1891 to 1908). Henri-Alexandre's brother, the future pastor Daniel Junod, would in his turn become a noteworthy palaeontologist. A. Grandjean, *La Mission romande* (Lausanne, 1917), pp. 268–9, 311–12; Dubois, *Naturalistes neuchâtelois*, p. 108. For a parallel situation with comparable outcomes, see J. Clifford, *Person and Myth: Maurice Leenhardt in the Melanesian world* (Berkeley and Los Angeles, 1982), pp. 13–15.
20 C. Dufour and J.-P. Haenni, *Musée d'histoire naturelle de Neuchâtel* (Hauterive, Neuchâtel, 1985), p. 46; Dubois, *Naturalistes neuchâtelois*, p. 51.
21 H.-A. Junod (étudiant), 'Les États de larve et de nymphe de l'hyponomeuta stannellus (Thunberg)', *BSNSN*, 14 (1884), 1–9.
22 The essays were published between 1880 and 1887. The praise came from 'J.C.', 'Chronique de Neuchâtel', *Revue de Belles-Lettres* 11:6 (1883), 219.
23 H.-A. Junod, 'Sur quelques larves inédites de Rhopalocères', *BSNSN*, 20 (1891–92), 18; 'La Faune entomologique du Delagoa – lèpidoptères', *BSNSN*, 27 (1898–99), 10; 'La Faune entomologique du Delagoa – coléoptères', *Bulletin de la Société Vaudoise des Sciences Naturelles* (henceforth *BSVSN*), 35 (1899), 162.
24 J. Naef, 'La Botanique', in J. Trembley (ed), *Les Savants genevois dans l'Europe intellectuelle: du XVIIe au milieu du XIXe siècle* (Geneva, 1987), pp. 360–7.
25 A. Forel, *Mémoires* (Neuchâtel, 1941), pp. 63, 164, 183ff.
26 This was a rare collection of butterflies. Ruth had married the widowed missionary pioneer Paul Berthoud in 1883 and returned with him to the Transvaal. Museum of Natural History, Neuchâtel, annual reports for 1884, 1886, 1889; *Nouvelles de nos missionnaires*, 1:9 (1886), 9.
27 E. Rambert, *La Société vaudoise des sciences naturelles* (Lausanne, 1876), pp. 25–6. See also W.-L. Distant, *A Naturalist in the Transvaal* (London, 1892), p. 124; D. E. Allen, *The Naturalist in Britain: A social history* (London, 1976), pp. 179–93.
28 Naef, 'La Botanique,' p. 364.
29 *Bulletin Missionaire de l'Eglise Libre du Canton de Vaud*, 85:7 (August 1889), 285.
30 SMA 503 Junod to secretary, 5 July 1889.
31 SMA 503 Junod to secretary, 30 October 1889.
32 G. Bertoloni, 'Illustratio rerum naturalium Mozambici', in *Novi Commentari Academiae Scientiarum Instituti Bononiensis* (Bologna), X (1849); W. C. H. Peters, *Naturwissenchaftliche Reise nach Mossambique in den Jahre 1842–1848* (Berlin, 1861, 1864, 1868).
33 When J. J. Monteiro, a British engineer, served as the Cape government's labour agent at Lourenço Marques from July 1876 until his death in February 1878, Rose helped him send dried plants, collected in the region, to Kew Gardens. She returned

in the late 1880s to complete his botanical work and collect saleable insects. Junod felt her *Delagoa Bay, its Natives and Natural History* (London, 1891) contained 'much interesting information,' although 'without claiming any great scientific accuracy,' H.-A. Junod, *Life of a South African Tribe* (London 2nd edn, 1927) (henceforth *LSAT*) II, pp. 147–8, n. 1. He was more critical of her work in a private letter to Eugène Autran, the conservator of the Boissier Herbarium, 19 May 1892 in CBG.HB.

34 H.-A. Junod, 'Le Climat de la baie de Delagoa', *BSNSN*, 25 (1896–97), 77–8.
35 These included the collections of Boivin, Gueinzius and Krauss, as well as the herbarium of Pierre Verreaux (1807–73). An associate of Andrew Smith at the Cape, Verreaux became a celebrated dealer in natural history specimens on his return to Paris. H. M. Burdet and A. Chapin, 'Les Herbiers de Genève', *Webbia*, 48 (1993), 238–9.
36 See Barbey's letter to Harry Bolus, February 1890 in University of Cape Town, Bolus Collection 234. Barbey had asked Paul Berthoud to collect plants in the northern Transvaal just as the region was being scoured by the German botanist R. Rehmann. See his *Polypetalae Rehmannianae* (1887–88).
37 CBG.HB Junod to Barbey, 30 October 1889; SMA 503 Junod to secretary, 30 October 1889.
38 SMA 503 Junod to secretary, Mission council, 30 October 1889; H.-A. Junod, 'Correspondences: de Rikatla à Marakouène', *Bulletin de la Société Neuchâteloise de Geographie* (henceforth *BSNG*), 6 (1891), 320; H. Schinz and H.-A. Junod, 'Zur Kenntnis der Pflanzenwelt der Delago-Bay', *Bulletin de l'Herbier Boissier*, 7:2 (1899). CBG.HB 'A propos de l'herbier de Shilouvane, apporté en 1893 [sic 1903] par Mr Henri-Alexandre Junod'.
39 Junod, 'La Faune entomologique – lépidoptères', pp. 186, 189; *Nouvelles de nos missionnaires*, 15:1 (1892), 11.
40 CBG.HB Junod to Barbey, 1 September 1891; *Nouvelles de nos missionnaires*, 15:1 (1892).
41 CBG.HB Junod to Barbey, 25 July and 4 October 1891. I have been unable to find this species as it does not conform to the rules of the International Code for Botanical Nomenclature.
42 CBG.HB Junod to E. Autran, 15 February and 5 September 1892; Junod had in fact learned to press and dry plants from a Paris Geographical Society publication for travellers.
43 CBG.HB Junod to Autran, 19 May and 5 September 1892; note entitled 'Collections Junod à vendre', 15 March 1904.
44 CBG.HB 'A propos de l'herbier de Shilouvâne'; Junod, *LSAT*, II, p. 238.
45 CBG.HB Junod to Barbey, 15 January 1903 and 24 May 1906; Junod to Schinz, 20 October 1903; Junod to Beauverd, 2 March 1910.
46 CBG.HB Junod to Barbey, 25 February 1898. In 1891 F. R. R. Schlechter (1872–1925) embarked on extensive botanical voyages in southern Africa, including Mozambique. F. A. Mendonça, 'Botanical collectors in Mozambique', in A. Fernandes (ed.), *Comptes rendues de la IVe réunion plénière de l'association pour l'étude taxonomique de la flore d'Afrique tropicale* (Lisbon, 1962).
47 CBG.HB Junod to Maurice Barbey, 25 April 1919; CBG.HB 'Compte des plantes de l'Herbier sudafricain de H. A. Junod expédié à l'Herbier Boissier,' annexure III attached to Junod to Professor Chodat, 3 January 1921. See also Junod's original botanical register, lodged in the Boissier Herbarium.
48 See the collections of J. Thunberg and N. L. Burman, Allioni, Houttuyn, Van Royen and others.
49 *LSAT*, II, p. 147, n. 1.
50 Schinz and Junod, 'Zur Kenntnis der Pflanzenwelt', continued in *Mémoire de l'Herbier Boissier*, 10 (1900).
51 See H. Schinz, 'Beiträge zur Kenntnis der Afrikanischen Flora', *Bulletin de l'Herbier Boissier*, vols 4 (1896), 5:12 (1896) and 7:1 (1899). J. Burtt-Davy was less glowing in his praise. Although he recognised the pioneering work of Junod around

Shiluvane, he found his plant specimens 'often scrappy and unfit for determination'. Burtt-Davy, 'First annotated catalogue of the vascular plants of the Transvaal and Swaziland', in *Report of South African Association for the Advancement of Science* (henceforth *RSAAAS*) (1908), p. 232.
52 See the *Index Kewensis*, 2 vols (London, 1996 edn). Also the entry on Junod in M. Gunn and L. E. Codd (eds), *Botanical Exploration of Southern Africa: An illustrated history* (Cape Town, 1981), p. 203.
53 *LSAT*, I, p. 1.
54 Junod, 'Rikatla à Marakouène', p. 322.
55 Junod, 'La Faune entomologique – lépidoptères', pp. 180, 184, 200, 219, 224, 240.
56 In December 1891 Junod claimed to have amassed a collection of beetles and butterflies worth just under £200, CBG.HB Junod to Barbey, 2 December 1891. On his sales to Trimen, see South African Museum, Trimen letterbooks, 31 July 1890 to 17 February 1892. Junod's gifts and sales of butterflies were noted in the Annual Reports of the Natural History Museum, Neuchâtel for 1892, 1894 (SF50 for 'insects') and 1911. Anonymous, *Le Musée d'histoire naturelle* (Neuchâtel, 1899), p. 40 mentions receiving from Junod at Rikatla a 'rich collection' of butterflies.
57 Junod, 'La Faune entomologique – coléoptères', p. 132; Junod, 'La Faune entomologique – lépidoptères', p. 177; Junod, 'Rikatla à Marakouène', p. 323.
58 E. Bugnion, 'Remarques supplémentaires', following Junod, 'La Faune entomologique – coléoptères', p. 189.
59 H.-A. Junod and O. de Schulthess-Schindler, 'La Faune entomologique du Delagoa: *Orthoptères*', *BSVSN*, 132 (1899); H.-A. Junod and A. L. Montandon, 'La Faune entomologique du Delagoa: *Hémiptères*', *BSVSN*, 132 (1899); H.-A. Junod and A. de Schulthess-Schindler, 'La Faune entomologique du Delagoa: *Hyménoptères*', *BSVSN*, 133 (1899).
60 Griffini, 'Sui Cybiser raccolti dal Rev. Junod a Delagoa', *Boll. dei Musei di Turino*, XIII, N° 325 (1898); Dr Regimbart, 'Monogr. Gyrinidae', *Ann. Soc. Entomologique de France* (1883) (sic 1893?); Regimbart, *Dytiscidae et Gyrinidae d'Afrique et de Madagascar* (Brussels, 1895).
61 L. Péringuey, 'A descriptive catalogue of the coleoptera of South Africa. Pt II', *Transactions of the South African Philosophical Society* (henceforth *TSAPS*), 7 (1896), 113–480; L. Péringuey, 'Descriptive catalogue of the coleoptera of S.A. – part III', *TSAPS*, 10:1 (1897), 23; L. Péringuey, 'Fourth contribution to the S.A. coleopterous fauna', *TSAPS*, 6:2 (1892), 95–6. L. Péringuey, 'Some new coleoptera collected by Rev. Henri A. Junod at Shilouvane, near Leydsdorp, in the Transvaal', *Novit. Zool.*, 11 (1904), 448–50.
62 He envisaged reworking published articles into chapters on the region's climate, flora, butterflies, coleoptera, orthoptera and hemiptera. CBG.HB Junod to Autran, 19 February 1900.
63 These included three butterflies and two moths. See E. L. L. Pringle, G. A. Hening and J. B. Ball (eds), *Pennington's Butterflies of Southern Africa* (2nd edn, Johannesburg, 1994), p. 33. For some of the other insects, including ants and rove beetles, see *Bulletin of the American Museum of Natural History*, XLV, 1921–22; Péringuey, 'Fifth contribution to the S.A. coleopterous fauna', *TSAPS*, 6:2 (1892), 248, 326–7, 479; W. L. Distant, 'Descriptions of new species of Hemiptera-Heteroptera', *Annals and Magazine of Natural History*, 2 (1898); Junod, 'La Faune entomologique – lépidoptères', p. 233.
64 The molluscs were mainly *agathines*, *Livinhacia Kraussi* and *Eerope caffra*.
65 In this way, for instance, it was discovered that two of the reptiles sent from Lesotho in 1907 by Mlle Jacot represented new species.
66 Junod, 'Le Climat de la baie', p. 78.
67 *LSAT*, I, pp. 65–6; II, p. 332.
68 UNISA.JC 3.3 'Elias, un ancien de l'église africaine'; Junod, 'La Faune entomologique – lépidoptères', pp. 179, 223.
69 H.-A. Junod, 'Les Ba-Ronga: étude ethnographique sur les indigènes de la baie de Delagoa', *BSNG*, 10 (1898), 21; *LSAT*, II, pp. 344–5.

70 *LSAT*, II, pp. 81–3, 344.
71 *LSAT*, II, p. 344.
72 Junod, 'Rikatla à Marakouène', p. 320; *LSAT*, II, pp. 329, 332, 345, 589; Junod, 'Ba-Ronga', 22.
73 CBG.HB Junod to Barbey, 2 January 1892.
74 CBG.HB, 'Botanique indigène', attached to Junod to Barbey, 16 October 1891. *LSAT*, II, p. 328; H.-A. Junod, 'The best means of preserving the traditions and customs of the various SA native races', *RSAAAS* (1907), p. 149.
75 CBG.HB Junod to Barbey, 5 July 1893. Plants known to supply antidotes against migraines, gonorrhoea and other maladies were sent to Mr Chodat in Geneva for medical analysis, see Junod to Autran, 1 December 1896.
76 H.-A. Junod, 'Une promenade aux environs de Rikatla', in anonymous, *Chez les Gouamba: glanures dans le champ de la Mission romande* (Lausanne, n.d. approx 1893), p. 11.
77 UNISA.JC 6. Conferences: à Neuchâtel, 1897.
78 Junod, 'La Faune entomologique – lépidoptères', p. 178.
79 Bryce wrote a highly favourable review of 'Les Ba-Ronga' and put Junod in contact with Sir James Frazer. Through Frazer, Junod gained access to the leading publisher of anthropological works in the United Kingdom, Macmillan. See Frazer's correspondence with Junod in the Wren Library, Trinity College, Cambridge University.
80 J. Bryce, *Impressions of South Africa* (3rd edn, London, 1899), p. 17. The last representative of the species of Cape lion died in the mid-1860s, and the last quagga in the Amsterdam zoo in 1883. It is quite likely that Junod saw the stuffed examples of these animals in the natural history museums in Berlin and Paris.
81 Distant, *Naturalist in the Transvaal*, pp. 41, 124–5.
82 Junod, 'La Faune entomologique – lépidoptères', pp. 228, 242; *LSAT*, II, pp. 80–1; Junod 'Promenade aux environs de Rikatla'.
83 Junod, 'La Faune entomologique – lépidoptères', p. 232; *LSAT*, I, p. 65.
84 See M. E. Barber, 'On the peculiar colours of animals in relation to habits of life', *TSAPS*, 1:4 (1877–78), 27ff.
85 Agassiz held these views until his death in 1873. In 1908 Junod acknowledged, on receipt of Mrs Barbey's translation of a work by A. R. Wallace, that he found the theory of evolution to be 'juste et feconde'. CBG.HB Junod to Barbey, 9 October 1908.
86 *LSAT*, I, p. 7.
87 Junod, 'Ba-Ronga', p. 481.
88 *LSAT*, I, p. 11; *LSAT*, II, pp. 609, 616.
89 L. Agassiz, 'Sketch of the natural provinces of the animal world and their relation to the different types of man', in J. C. Nott and G. R. Giddon (eds), *Types of Mankind* (London, 1854); E. C. Agassiz, *Louis Agassiz, His life and correspondence* (Boston, 1885), vol. II, pp. 598ff.
90 Junod, 'Ba-Ronga', p. 480.
91 H.-A. Junod, *Grammaire Ronga suivie d'un manuel de conversation et d'un vocabulaire ronga-portugais-français-anglais pour exposer et illustrer les lois du Ronga, langage parlé par les indigènes de district de Lourenço-Marques* (Lausanne, 1896), p. 2.
92 Junod devoted the second volume of the first edition of *LSAT* (1913) to the 'Psychic Life' of the tribe. In the second edition (1927) he anglicised this term into 'Mental Life'.
93 Junod, 'The best means', pp. 151–2.
94 Junod, 'The best means', p. 153. He seemed to be unaware of how photographs were infused with meaning when selected, provided with captions or even tampered with. See P. Harries, 'Terrible truths: Swiss missionaries and the role of photography in the early ethnographic monograph.' Paper presented to the conference 'Encounters with Photography', Cape Town, July 1999.

95 *LSAT*, I, p. 11.
96 Junod, 'The best means', p. 153; *LSAT*, I, p. 1. Sir James Frazer and Lucien Lévy-Bruhl performed this function for Junod. See also Pittard, n. 108 below.
97 *LSAT*, I, p. 3.
98 His modesty was at least partly inspired by his need to solicit the patronage required to break into the world of 'anglo-saxon' anthropology. Lord James Bryce fulfilled this role for Junod, see n. 79, above.
99 *LSAT*, I, pp. 8–9.
100 H.-A. Junod, 'Some remarks on the folklore of the Ba-Thonga', *Folk-lore*, 14:2 (1903), 121, 123; *LSAT*, II, p. 329.
101 Junod, 'Ba-Ronga', pp. 7–8, 245, 247.
102 He never abandoned his belief that the evolutionist hypothesis was 'the best solution to many problems'. See H.-A. Junod, *Moeurs et coutumes des Bantous* (Paris, 1936), I, p. 256; H.-A. Junod, *Le Noir Africain. Comment faut-il le juger?* (Lausanne, 1931), pp. 5, 18–19.
103 Junod, 'Ba-Ronga', p. 114.
104 In 1898 Junod discounted the physical threat to the Thonga of the vices brought by industrialisation. Thirty years later he considered the extinction of 'the Black race,' and especially the Thonga tribe, a distinct possibility. See 'Ba-Ronga,' 7; *LSAT* II, pp. 629–30.
105 See Junod, 'Ba-Ronga', p. 116; *LSAT*, I, p. 11; *ibid.*, II, p. 632. Those who adopted this view of human species included both Creationists like Agassiz and convinced Darwinians like Carl Vogt and August Forel.
106 See *LSAT*, II, p. 596.
107 Junod, *Le Noir Africain*. Junod did, intriguingly, consign long passages on female genitalia to an appendix written in Latin. It is unclear whether his interest in this subject was occasioned by Darwin's ideas on the relationship between sexual selection and species differentiation. It might equally have been the product of the botanical and entomological practice of classifying specimens according to their sexual parts.
108 See E. Pittard, 'Contribution à l'étude anthropologique des BaRonga', *Bulletin de la Société Neuchâteloise de Geographie*, 26 (1917), 159. This article was based on Junod's measurement of twenty-two Rikatla students.
109 CBG.HB Junod, 'Botanique indigène'; Schinz and Junod, 'Zur Kenntnis der Pflanzenwelt', p. 888.
110 See Junod, 'Ba-Ronga', p. 16; CBG.HB Junod, 'Botanique indigène'.
111 See *LSAT*, II, pp. 328–32.
112 *LSAT*, II, pp. 344–5.
113 Junod, 'La Faune entomologique – coléoptères', p. 184; Junod, 'Ba-Ronga', pp. 419–20; *LSAT*, I, p. 65; *LSAT*, II, pp. 80, 83, 341–2.
114 *LSAT*, I, pp. 9, 166, 521.
115 Junod, 'The best means', p. 143; H.-A. Junod, 'The magic conception of nature amongst Bantus', *South African Journal of Science*, 17 (1920), 79.
116 A. Isaacman, (ed.), *The Life History of Raúl Honwana: An inside view of Mozambique from colonialism to independence, 1905–1975* (Boulder and London, 1988), p. 58.
117 See the reviews by E. Durkheim and M. Mauss of 'Les Ba-Ronga' in *L'Année Sociologique*, 3 (1898–99).
118 H. Kuper, 'Function, history, biography: Reflections on fifty years in the British anthropological tradition', in G. W. Stocking (ed.), *Functionalism Historicized: Essays on British social anthropology* (Madison, 1984), p. 195; Hammond-Tooke, *Imperfect Interpreters*, p. 199.
119 S. Dubow, *Scientific Racism in Modern South Africa* (Cambridge, 1995), p. 130; A. D. Spiegel, 'Towards an understanding of tradition', *Critique of Anthropology*, 9:1 (1989).
120 A. R. Radcliffe-Brown, preface to M. Fortes and E. E. Evans-Pritchard (eds), *African Political Systems* (London, 1940), p. xi.

121 E. R. Leach, *Rethinking Anthropology* (London, 1966), pp. 2–3.
122 A. Gupta and J. Ferguson, 'Discipline and practice: "The field" as site, method, and location in anthropology', in their (ed.), *Anthropological Locations: Boundaries and grounds of a field science* (Berkeley and Los Angeles, 1996), pp. 1, 20–1. For modern utilisations of Junod's work, see Luc de Heusch, 'The Thonga's goat', in his *Sacrifice in Africa: A structuralist approach* (Manchester, 1985); A. Kuper, 'Radcliffe-Brown, Junod and the mother's brother in South Africa', in his *South Africa and the Anthropologist* (London, 1987); P. Harries, *Work, Culture and Identity: Migrant labour in Mozambique and South Africa, c. 1860–1910* (Portsmouth, NH, 1994).

CHAPTER TWO

Making canes credible in colonial Mauritius[1]
William K. Storey

Mauritius rarely attracts the attention of southern African historians because the island, located in the south-west Indian Ocean, may or may not be classified as part of southern Africa. There are some connections: Cape merchants have been trading with Mauritius ever since the seventeenth century. There are also some similarities: Mauritius resembles Natal because, during the nineteenth century, Mauritian and Natalian planters employed indentured labourers from India. Some Natal planters can also trace their ancestry to Mauritius. There are also some broader comparisons that might be made: in Mauritius, as in much of southern Africa, large-scale capitalists relied on densely populated rural areas to provide inexpensive labour. But to say that Mauritius is a 'part' of southern Africa may be to stretch arbitrary regional boundaries beyond recognition. Most obviously, the geographic expanse of southern Africa contrasts with the insularity of Mauritius. More to the point, few Mauritians consider themselves to be African, even Mauritians of African ancestry.

Then why should any student of southern African history care about the island's history, let alone the history of Mauritian agricultural science, the subject of this chapter? Evidence about the history of science on the island raises interesting questions. Some previous historians have argued that European experts imposed the practices of anthropology, ecology, medicine and religion onto Europe's colonial subjects. Yet the history of agricultural science in Mauritius shows that new kinds of colonial science were the subjects of intense local debates. During these debates, colonial subjects came to influence the production and distribution of knowledge. This was not a part of any state programme to get plantation owners and peasant farmers more involved in the work of research institutions; it was part of a long-term process in which Mauritians creolised colonial politics and colonial science. In Mauritius, lay agricultural knowledge became a part of colonial scientific knowledge. This story resembles Carolyn Hamilton's work on Shaka Zulu in that Hamilton shows how the myths associated with Shaka can be traced to colonial discourses which, in turn, originated in contemporary African and European accounts.[2] While the story of Shaka is rather different from the story of Mauritian agriculture, both stories call attention to the involvement of colonial subjects in the production of colonial knowledge.

Scientific knowledge was not just important for farmers and scientists. Colonial scientific practices were fundamental to colonial political practices, and vice versa, even as science and politics underwent fundamental transformations. The case of Mauritius may raise new questions for southern African historians interested in the politics of scientific research. In a recent study, Saul Dubow has shown the ways in which South African scientists elaborated on and contributed to European racial science: 'metropolitan' European science did not simply determine 'peripheral' thinking in South Africa; rather, there was a complex interplay between thinkers in both places.[3] This chapter develops a similar argument about Mauritius, but I shall pay more attention to debates over scientific expertise. Which scientists are the most believable: researchers with internationally recognised credentials, or 'laymen' who have hands-on or 'tacit' knowledge? How does epistemological credibility pertain to social standing, especially in a rapidly evolving colonial society with sharply differentiated classes and identities? The answers to these questions may help to inform debates about reformulating scientific practices under the new regimes of southern Africa.

Mauritius was at first uninhabited, and it was not until the seventeenth century that the Dutch East India Company made the first attempts at colonisation. The Dutch colony did not last long: remoteness from markets and administrative centres as well as problems with droughts and hurricanes persuaded the Dutch to abandon the island. The French took over in 1715, and over the course of the eighteenth century they transformed the island into an important trading post and refreshment station. A three-tiered society began to emerge, comprising a thin veneer of Europeans and mixed-race freedmen who benefited from the work of a slave population that originated mainly in Africa but also in Asia.

In 1810, the British captured Mauritius and converted the island from a maritime base into an agricultural colony. Sugar cane grew best of all commercial crops because of its resistance to damage by hurricanes. By the 1820s, the island's planters were growing sugar cane to the exclusion of almost everything else. As cane cultivation spread, the minority community of Franco-Mauritians converted their domestic and maritime slaves over to cane-field labour and smuggled new slaves to the island in defiance of Britain's ban on slave trading. When Mauritian slaves became free in 1835, the sugar estates used compensation payments from the British government to import indentured labourers from India. These labourers were a mix of Hindus and Muslims from all regions of India who spoke many different languages and dialects. As Indo-Mauritians they became the majority community in Mauritius during the second half of the nineteenth century.

At this time, world sugar prices began to decline steadily. This caused the Mauritian sugar industry to rationalise production, a process that introduced significant economic, social and technical changes. Most sugar estates either consolidated their lands into larger company holdings or sold parcels of land to Indo-Mauritians. By the first decade of the twentieth century, approximately two-thirds of the island's sugar canes were still grown on estates of over four hundred hectares, most of which were owned by Franco-Mauritians. The remaining one-third of the canes was mostly grown on Indo-Mauritian peasant plots that usually did not exceed four hectares.[4]

Social and economic stratification could be seen in political representation. At first, the British imposed the absolute rule of a governor and a small council of officials, but soon they found that the island ran most smoothly when they incorporated the Franco-Mauritian elite unofficially within the decision-making process. This was the first indication that the Mauritians could creolise the colonial state; even though Mauritians did not hold ultimate political authority, they were able to participate in the government and shape its practices. In 1886 a sympathetic governor granted Mauritius a partly representative government, for which only the wealthiest two per cent of the island's population could vote. This system excluded all but a few Indo-Mauritians and Creoles, the term used to describe the descendants of the freedmen and slaves. Between 1937 and 1957, the colonial government acceded to popular pressure and introduced gradual democratic reforms. In 1968 Britain granted independence to the island.[5]

Under British rule, Mauritius became a colonial society in the purest sense of the word: everyone was descended from migrants, everyone was nominally ruled by a remote sovereign and everyone depended on sugar cane. Sugar cane farmers, in turn, depended heavily on sugar cane researchers. Research began during the eighteenth century, when French plantation owners imported new cane varieties from overseas. Working together with local naturalists, planters used empirical methods to select the canes that were best adapted to local circumstances. These efforts continued even after the imposition of British rule in 1810, but cane imports were not enough to protect the colony's monocrop economy.

In the 1860s, the principal varieties of the sugar cane plant started to succumb to diseases, a situation that worried the plantation owners' main organisation, the Chamber of Agriculture, so much that the organisation asked for help from the British government. In 1868, the governor of Mauritius sent a botanist from the Royal Botanic Gardens, Pamplemousses, to the Pacific. The botanist, Dr Charles Meller, collected new sugar cane varieties over the course of several months. He died from malaria, but managed to send his canes back to Pamplemousses, where his assistant, John Horne, planted the canes.[6]

Even with such dedicated government services, the members of the Chamber were not happy to designate the British botanists as their official cane collectors. Several members questioned Meller and Horne's ability to identify and cultivate canes properly, even though both 'gardeners' were supported by Kew Gardens, the central imperial botanical institute in England. One Mauritian planter cautioned that while the gardeners may have been competent botanists, 'well-versed in the means of flowers and trees', they might have needed 'practical advice in cane culture'. The president of the Chamber added that he thought it possible to be a good botanist and a poor cane cultivator.[7]

The Chamber addressed the possible incompetence of the botanists by appointing a committee to supervise the cane collection. They asked the governor for permission to visit the collection periodically, and the governor granted their wish. He instructed the director of the gardens to allow the Chamber's cane committee 'to have access at any time to these canes, and to receive any advice which the committee may offer you on the subject of their cultivation and propagation'.[8] Nobody seems to have doubted that the Chamber had a right to supervise the work of a colonial state garden. Politically speaking, government cooperation with planters reflected the broader influence that the planters had over the colonial state.

For the next thirty years, the cane committee of Franco-Mauritian planters visited the Pamplemousses Gardens and supervised the work of John Horne and the other British gardeners on the sugar cane collection. The collection grew to about 200 different varieties, one of which, the Tanna cane, became practically the only cane cultivated in Mauritius between the 1890s and the 1930s. But even after the successful cultivation of canes at Pamplemousses, one senior member of the Chamber stated that the British staff were 'well-meaning' but still 'a bit foreign to cane culture'.[9] Being held in such low esteem made it difficult for British botanists to persuade Franco-Mauritian planters of the superiority of new canes. The planters may have been laymen; John Horne once described a delegation from the Chamber as 'a set of ignoramuses who meddle with matters they know not of'.[10] Even so, the planters were still able to use their political influence over the governor to establish the social and rhetorical norms of experimentation.

This point about the close relations between scientific and political practice echoes several studies of early modern European science. In that field, several historians, such as Mario Biagioli, Peter Dear, Steven Shapin and Simon Schaffer have written about the relationship between the social status of scientists, their rhetorical and social practices, and their epistemological credibility. According to these scholars, scientists design experiments so that they may be verified; they select appropriate witnesses to approve their experimental practices; and they publish their results so that they may be replicated. Credibility depends to a great extent on the ways in which scientists report their experiments. In particular, credibility depends upon scientists cultivating politically appropriate relationships. Science and politics are not two separate ways to produce order in nature and society; rather, they are often so intertwined that it is difficult to disentangle them.[11]

The close relationship between science and politics in colonial Mauritius appeared even more evident when planters became interested in founding a new research institutution. As early as 1877, the Chamber's president, Virgile Naz, proposed that the Chamber create an agricultural research centre. Naz was a Creole lawyer and landowner who was assimilated into the elite. Even so, the issue of who might have access to the institution complicated and stalled the project. Naz, who was also an amateur agronomist, suggested that the Chamber establish a 'section agronomique' to keep track of agricultural experimentation on the island.[12]

Some prominent members rejected Naz's idea vehemently. One critic argued that growers who did not belong to the Chamber would not be paying for the research at Naz's centre, and therefore should not benefit from its findings, a suggestion that would have been difficult to enforce. According to him, the general public ought not to have access to privately funded research. Besides, even though a substantial number of Mauritians had an interest in research, a minority of the sugar estate owners and staff members were competent to carry out reliable experiments. The Chamber tabled Naz's idea and explored other avenues, including a proposal from a private nurseryman to grow and distribute new cane varieties. In 1884, the president also suggested that the government subsidise a Chamber scheme to teach agronomy, but nothing came of these efforts.[13] Naz's only accomplishment was to have broached the subject of independent research; other matters, such as price declines and forest regulations, absorbed the sugar industry's attention.

Shortly thereafter, a significant discovery in sugar cane botany recaptured the attention of the sugar industry. For centuries, Europeans had thought that sugar cane was not fertile, and could only be propagated by cuttings. The principal cane varieties in cultivation around the world, notably the Creole and Bourbon, were indeed not fertile, but during the 1850s and 1860s, when growers around the world began to introduce new canes from the Pacific, some began to notice seedlings in their fields.[14] At first, Mauritians were slow to realise the potential of seedling canes. As early as 1860, an estate manager proposed producing seedlings, but Edmond Icery, the Chamber's president, warned him that if he tried, he would be ridiculed. In 1871, another estate owner reported to Icery that he noticed seedlings in one of his fields. Neighbouring estate owners verified his claim, but he could not produce a germinated seed as evidence, much to Icery's chagrin. Mauritian naturalists mistakenly confirmed (with their correspondents in Réunion and Kew) that cane seeds did not exist.[15]

In spite of this apparently unwarranted discouragement, Icery continued his observations. In 1880, he recounted that he noticed some seedling canes himself, but he never could find a remnant of the seed in the roots. He examined a cane arrow under a microscope, observing correctly that out of the tens of thousands of minute flowers, the male flowers were closed. He deduced correctly that the cane was sterile. He added that the flower could perhaps be pollinated artificially if one opened up the male flowers.[16] He was actually wrong about this, but his efforts did show that colonial planters without metropolitan botanical qualifications were nevertheless able to conduct some of the most significant research.

Even without qualifications, Mauritian planters still had the power to supervise the old colonial botanic garden. Nevertheless, more of them began to want a new kind of research institution. This was at least partly because of developments outside of Mauritius. In the United States, Europe and some of Europe's colonies, agricultural research was moving out of the fields and gardens and into research laboratories.[17] Yet even as the new experimental approach succeeded, its spread was not simply a case of new methods and discoveries forcing farmers to re-evaluate their approaches to the management of natural resources. Farmers had to be persuaded that new procedures and institutions really worked. Scientists did so by building networks of people, organisms and inanimate objects to support their claims.[18] Changes in the production and dissemination of new sugar cane plants can be understood in the context of global changes in colonial research institutions. They may also be understood in terms of demonstrable results. But they must also be understood in the context of local economic, political and social relations.

It was not until 1885, when William Newton assumed the Chamber's presidency, that any serious action took place in building an experiment station and thereby improving Mauritian agricultural research. Newton was an aggressive Creole lawyer who had represented the Chamber before a critical commission of enquiry in 1874. Since then he had acted occasionally as the Chamber's spokesman. In 1885, he recapitulated the ideas of the Chamber in a pamphlet entitled 'The Sugar Crisis'. There, Newton complained bitterly about the European bounties on beet sugar and argued that sugar canes enjoyed natural advantages over sugar beets. Newton suggested that research and teaching could improve the competitiveness of Mauritian cane, adding, 'What we want is a few less planters and a few more agriculturists.' To achieve this objective, he proposed the creation of an agronomical research station and training centre.[19]

The Chamber was slow to realise Newton's goal because it proved difficult to negotiate the respective roles of the private and public sectors. Even though the proposed research station was to be private, the Chamber still sought government cooperation, particularly in finance. Governor Pope Hennessy and the Council of Government endorsed Newton's pamphlet and forwarded it to the Colonial Office in London. The Chamber and the colonial government agreed that in order to attract a qualified scientist to head a 'station agronomique', they needed to pay the director Rs.10,000, double John Horne's salary at Pamplemousses. The Colonial Office in turn consulted with Kew Gardens, which approved the idea of creating a research station, while criticising Mauritian reliance on sugar cane.[20] But while these proposals were circulating, Pope Hennessy became embroiled in other disputes with the colonial government, pushing ideas to improve sugar production off the political centre stage.

Newton persisted, using his friendship with Pope Hennessy to press his ideas. In 1886, the Colonial Office recalled Pope Hennessy to question the governor about allegations that he had meddled in the 1885 elections. Newton accompanied Pope Hennessy to London as his legal adviser. During the voyage, Newton made the acquaintance of a French agricultural chemist who helped him to sketch out a plan for the 'station agronomique'. His shipmate also recommended that Newton hire Philippe Bonâme, a well-known chemist who was in charge of the agricultural research station in Guadeloupe, to manage the future Mauritian centre. Upon arriving in London, Newton met with the Secretary of State for the Colonies and convinced him to approve the creation of a 'station agronomique'.[21]

Newton's meeting with the secretary of state increased the pressure on the government of Mauritius, but so did the first experimental production of cane seedlings in Java and Barbados. Apparently, while other regions made progress towards breeding their own canes rather than importing them, Mauritians were spending most of their time bickering.[22] When Kew sent the first seedlings from Barbados, Mauritians finally realised that a new era was dawning in their understanding of the sugar cane plant, and this inspired widespread amateur breeding experiments. John Horne asked Kew to 'Kindly send me all the papers, your own and those of others, you can get on growing sugar canes from seed. We are all agog on the subject just now.'[23] However, he knew from his original failure to germinate the seeds that cane breeding would be a formidable task. Horne predicted that, 'Raising cane from seed to get improved varieties will be a long and tedious affair, and there will be many disappointments before a really good, hardy sugar-yielding variety will be obtained.'[24]

The Mauritians did, in fact, have many disappointing canes ahead of them, but at the time the planters only knew that Horne was giving unpalatable advice. Horne's pessimism carried no sway. In September 1887, the Chamber formed a Station Agronomique Committee to prod the government into action. In May 1888, the committee reported that the island's sugar cane growers should bear the costs of the Station Agronomique since they benefited from it most directly. The easiest way to raise the funds to provide for the station, according to the committee, was to ask the government to introduce a light tax on sugar exports. The producers would pay to support the station according to the proportion of sugar they produced. The committee estimated that it would cost Rs.60,000 to establish the station and Rs.34,000 to maintain it each year, including Rs.10,000 to pay the director's salary. The starting and annual expenses could be met from a tax of two cents on every fifty kilograms of sugar exported, which was enough to furnish Rs.40,000 each year. The Chamber expected the governor to donate some land at Réduit for the station, thereby reducing its expenses even further.[25]

The proposal seemed straightforward, but there was one tricky question: should representatives of private industry or government serve on the board that was to oversee the station's work? Not surprisingly, the Chamber's committee recommended themselves for the job, arguing that if the sugar industry were to pay for the station, the

Chamber should form a committee to ensure it carried out its mission. They were adamant that the station should not be made a government department.[26]

By then, members of the Chamber were accustomed to having a strong voice in local governance, but they should have known that their request for government assistance in raising revenues would cause the government to demand a voice in setting the research agenda. At the governor's request, the Council of Government appointed its own Agronomic Station Committee to study the question.[27] It included many of the same members of the Chamber who had been working to create a research station for years, including Naz, Newton and the chairman, Henri Leclézio, who owned an estate in Moka but who was also a rising power in island politics. The presence of several colonial officials ensured that the committee's proceedings were not completely redundant, but in the end their recommendations served the interests of the sugar industry. The committee heard testimony from six factory chemists, a nurseryman and John Horne of Pamplemousses, allowing for a more informed debate on the mission of the station.[28]

Disagreements about the station's institutional arrangements continued. This caused some planters to take matters into their own hands and to explore alternative methods to obtain new canes. Flushed with one estate's success, the Chamber's president made the sanguine suggestion that Mauritius might match the sugar beet breeders, who had doubled the sucrose content of their plant. Following Mauritian standard operating procedures, one owner persuaded the Chamber to name a committee charged with determining and disseminating the best methods for raising canes from seeds. In 1891, interest in amateur cane breeding peaked, thanks to the Chamber's encouragement, but shortly thereafter it waned when people began to appreciate the difficulties involved. The Chamber offered a prize of Rs.1,000 to the grower with the best seedling canes and published instructions on how to germinate cane seeds. Perhaps a dozen individuals attempted to grow seedlings, and the award went to Georges Perromat, the manager of Clemencia estate in Flacq, who produced 287. Perromat invested his winnings in a nursery where he specialised in producing seedlings for the sugar industry. In 1893, he sold his collection of 500,000 seedlings to the Mauritius Estates and Assets Company, which planted them at Beau Champ estate in Flacq. Perromat did not take a systematic approach to cane breeding; by 1894 the company realised that the vast majority of his seedlings were worthless and offered them to anyone who might be interested. Nevertheless, Perromat continued with his work and several of his varieties were used as commercial canes, accounting for twenty per cent of the area under cane in 1915.[29]

Even as Perromat and other Mauritians bred seedlings, negotiations continued over the creation of the interdisciplinary research station. Finally, the government and the Chamber settled on the station's research agenda. The sugar industry was coming to recognise that an interdisciplinary scientific approach to agriculture would produce greater profits. Most of the experts testifying before the committee agreed that the station should work on selecting new cane varieties while also addressing soils, manures, plant diseases and methods of cultivation. John Horne continued to make himself unpopular by arguing for research on crops other than sugar cane, as well as on animal husbandry. Horne also suggested establishing three stations, each one in a different climatic zone, but no one took this sensible suggestion seriously. In the end, the committee's recommendations for the station's agenda reflected Newton's original ideas, which he himself had derived from correspondence with France's Station Agronomique de l'Est at Nancy.[30]

The new agenda represented a major shift away from the prior work of the Pamplemousses Gardens, where the staff worked mainly on collecting and cultivating new varieties. In 1891, the Chamber's secretary spoke of the economic dangers of Mauritius staying 'mired in the ideas of yesterday', when agricultural research was becoming increasingly oriented to laboratory experimentation.[31] The Chamber sought instead to emulate the interdisciplinary laboratory research currently in vogue in Europe. During the same year, the Chamber's president called upon his estate-owning colleagues to 'arm' themselves with the 'weapons' of science to fight the 'common enemy, beet root sugar'. He berated the Chamber for taking so long to create a research station, when 'the lands are exhausted, the yields diminishing', and the popular Port Mackay cane variety was degenerating.[32]

Throughout the negotiations, the Chamber of Agriculture's representatives on the committee sought direct control over the new station's research agenda. The Chamber proposed 'looking for no external aid from government, but relying upon our own exertions'.[33] Even so, the government's ability to raise scarce capital through taxation prolonged the collaboration between the state and the sugar industry in scientific matters. While the Chamber accepted some of the Nancy station's suggestions in scientific matters, French recommendations for the management of the institution itself proved unpalatable. Nancy urged that the secretary of state appoint a director who would be financially independent of the producers. The Chamber did not consider it 'advisable' to make the director dependent on the government, citing their experiences with the director of the Pamplemousses Gardens. The Chamber 'was of opinion that the station should be placed under their control, inasmuch as they claim to represent the Agricultural Body that would be called upon to pay the cost of the erection and upkeep of the station'.[34]

The Chamber claimed the virtual representation of the entire planting community, in much the same way as the Westminster Parliament claimed to represent all of Britain before the reforms of 1832. In other words, they claimed to represent everyone, even though only a tiny portion of the population had voted for them. In Mauritius, the emerging classes of peasants had no voice in government. Had anyone bothered to consult them about the creation of the Station Agronomique, they might have disagreed with the Chamber's position. After all, the Chamber was proposing that all sugar producers, large and small, would pay the export duty to support the station, but only the estate owners who belonged to the Chamber would control the production and distribution of its scientific results. The Chamber even had the audacity to ask the government to give it Rs.5,000 of the new tax revenues to defray its own expenses.

The Chamber's discussions with the government about creating the Station Agronomique, and the work of the committees involved, all produced an entirely unsurprising result: the Chamber got almost everything that it wanted. The export tax would pay for the annual operating expenses and the salaries of the staff. The government forwarded the station Rs.30,000 to construct buildings and Rs.20,000 to equip the laboratories, sums that the export tax would pay back gradually. The station was small, but government backing gave it some potential.[35]

The Chamber also received most of the power to supervise the Station Agronomique. This was an important step, because the production of experimental knowledge is not simply a matter of creating new facts. The credibility of new discoveries rests heavily upon the social and political conventions for regulating and disseminating the new knowledge.[36] This is what the Chamber's supervision of the Pamplemousses canes had already demonstrated. Originally, the Chamber proposed the creation of an oversight board with six members, to be called the Station Agronomique Committee. It recommended that three elected members and one non-official appointed member of the Council of Government sit on the committee, together with the presidents of the Chamber of Agriculture and the Royal Society of Arts and Sciences. Most of the individuals holding these offices were likely to be members of the Chamber. The Chamber requested that the committee be able to elect its own chairman as well.[37] The Chamber's oversight committee would verify the Station Agronomique's experiments less directly than the old Sugar Cane Committee that visited the state gardens at Pamplemousses. This may have reflected the more extensive bureaucratic influence the Chamber would have in the new Station Agronomique.

The British colonial government of Mauritius granted the Chamber most of its wishes, but a struggle took place over whether ultimate control of the station should rest with the government or the Chamber. In February 1891, the Colonial Office instructed the governor to endorse the Station Agronomique Committee's recommendations, but to resist its efforts to appoint and supervise the director.[38] In July 1892 Henri Leclézio, chairman of the Station Agronomique Committee, requested that the Colonial Office reconsider. He argued that committee oversight of the station's director would be important, because planters understood sugar cane agronomy better than government officials. He also argued 'that as the sums spent on the station were to be recouped by means of a special tax on sugar it was but just that the planters should have a voice in the management of that establishment'. Leclézio proposed that the governor and secretary of state appoint the committee from the ranks of prominent planters. He reassured the Colonial Office that the colonial government's auditor general would review the station's expenditures because the funds passed through the treasury. The governor endorsed these proposals. He also agreed with the committee's recommendation to appoint Philippe Bonâme, a French chemist completely outside Kew's network of botanists, as the new director.[39] Nobody paused to consider whether the sugar industry's representatives in the Chamber and on the committee might only really represent the large estates with factories. Even the Colonial Office in London weighed the Mauritian sugar barons' demands favourably against the established pattern of consulting with Kew about botanical institutions. In a minute attached to Leclézio's report and initialled by the secretary of state, Lord Knutsford, two Colonial Office bureaucrats wrote:

> [1st]: I think we may accede to all these proposals [by Leclézio]; it is so purely a matter of local interest that it is best not to interfere with local wishes, especially as the Director is only to be appointed on a five years agreement. I would not therefore consult Kew as might otherwise be desirable, but assent to what is proposed...
>
> [2nd]: In answering the despatch say that the S of State will not object to the management of the Station Agronomique being placed under the control of a Committee as proposed but ask why the Committee were appointed by the Council of Govt and not by the Governor.[40]

The Colonial Office's recommendations on the composition of the Station Agronomique Committee were followed by more confusing negotiations in Mauritius. Technically the Colonial Office had the last word, but practically the colonial government of Mauritius could accomplish little without the support of the sugar industry's elite. The Council of Government, including some members of the Chamber, passed a resolution allowing the governor to appoint the committee. The governor allowed the Council to vote on his appointees anyway. The governor also appointed Philippe Bonâme as director of the station in consultation with the committee. The officials of the Colonial Office felt uneasy about the governor's actions, but resigned themselves to a policy of non-intervention.[41] In June 1893, the Station Agronomique opened officially on land adjacent to the governor's mansion at Réduit.

The Station Agronomique accomplished much under the direction of Bonâme. He studied all aspects of the sugar industry, but mainly he experimented on different soils and fertilisers to see how they affected sugar production. Bonâme also researched the cultivation of other plants, as well as economics, entomology, pathology, implemental tillage, animal husbandry and factory chemistry. In 1901 alone, in addition to analysing hundreds of sugar cane samples, he investigated irrigation practices, looked into feeding cows a mixture of bagasse and molasses to get them to produce more manure, and weighed the benefits of unleashing imported owls on the island's rat population.[42] A small interdisciplinary research station could not have been expected to accomplish more. However, he failed to produce the results that the Mauritian sugar estates desired.

The choice of Philippe Bonâme as director caused some problems and, partly because of him, the station never lived up to its potential. Bonâme was happiest working in his laboratory. He was an able scientist, but his shyness prevented him from lobbying effectively for the station among his loquacious patrons. Throughout the station's twenty-year existence, the Chamber's minutes show that he attended meetings often but spoke rarely. When the members of the Chamber tried to draw him out, as they did in October 1908 when they invited him to a gathering of estate owners and staff members in Flacq, he gave a long, technical discourse on agricultural science that must have bored his amateur audience.[43] Someone with greater social and political skills might have been able to make the sugar industry more aware of the importance of basic research on the sugar cane plant and thereby increase his institution's budget and staff. However, Bonâme failed to create the kind of political and social network needed to sustain a new experimental institution in Mauritius.

Bonâme's efforts came up short in the critical area of providing new cane varieties for the sugar industry. Anyone with a good knowledge of cane cultivation and elementary botany could have participated in the old system of plant transfers, which involved comparing the performance of imported varieties. Bonâme was perfectly capable of this, even though his formal qualifications were in chemistry. Bonâme was also capable of producing and selecting seedling canes, as were many of the amateur scientists among the planting elite. But Bonâme did not do any better than the amateurs in providing new seedling cane varieties suited to the Mauritian sugar industry. Almost every year he used systematic methods to produce more than a thousand seedlings, but none of them were as good as the amateur Perromat's seedlings. In fact, by 1915, five of Perromat's varieties occupied 20 per cent of the area under canes, while Bonâme's canes were nowhere to be seen.[44] Bonâme's slow start in producing seedling canes, coupled with the fact that proper cane selection trials often last for many years, wore the sugar industry's patience thin. Bonâme was trapped between high expectations and modest capabilities. The Station Agronomique failed as an institution because its director, while producing competent scientific research, did not produce better canes. At the same time, he did not relate his findings well to the island's farmers.

As early as the 1890s, the Chamber became frustrated with Bonâme's first attempts at seedling cultivation, and they fell back on the Pamplemousses Gardens' cane collection to obtain new varieties. The Gardens' main success was in distributing the Tanna cane, which became the principal variety cultivated on the island until the mid-1930s.[45] The staff also attempted breeding experiments unsuccessfully and assisted local growers in identifying new seedlings and mutations. Pamplemousses also imported and distributed new seedling varieties from the West Indies.[46] Recognising the continued importance of the Gardens, the Chamber engaged in a long-running dispute with the government over the petty details of cane distribution, which in the end confirmed the Chamber's proprietary rights over all the canes.[47] Until 1898, the Chamber continued to use Pamplemousses as a source of canes. In 1901, it used the Gardens' collection only as a reserve in case some varieties failed in the Station Agronomique's cooler climate.

The station gradually took charge of producing and distributing new cane varieties for the sugar industry. Bonâme started a repository of living cane varieties at Réduit with the help of the Chamber, which provided cuttings of the fifty-one different varieties remaining at Pamplemousses.[48] The station also began receiving all newly imported varieties, at first through Pamplemousses and later directly from foreign exchange partners. During the mid-1890s, Bonâme began scientific breeding experiments, attempting to follow the methods then employed in Barbados and Java. By the early 1900s, he had the sole responsibility for providing canes in Mauritius. The Chamber simply trusted him more than the staff at Pamplemousses, whom one member derided as 'well-meaning' but still, after three decades of cane-growing, 'a bit foreign to cane culture'.[49] They believed in Bonâme because of his superior credentials and also because of the station's institutional ties to the Chamber.

Soon enough, Bonâme himself proved 'a bit foreign' to cane breeding. He used wind pollination to produce seeds, the most common method in the world at that time. Then he selected the best new seedling varieties after several years of field trials. He took the insightful step of planting his seedlings at Pamplemousses as well as Réduit and shared some with estates around the island to see if the plants grew better in particular locales.[50] However, Bonâme's studies were just as flawed as those of the cane breeders in most other regions: his field methods did not allow him to keep track of which parent canes produced better offspring. The consummate laboratory scientist, Bonâme performed extensive chemical analyses on his seedlings, comparing their sucrose content and other qualities with the imported canes. Bonâme's preferences reflected his training as an agricultural chemist. Even though Bonâme was French, giving him some things in common with the Franco-Mauritian owners of the sugar estates, his intense reliance on chemistry distanced him from the sugar-growers. While European chemists were expected to conduct rigorous investigations in the laboratory, the credibility of Mauritian sugar cane scientists rested on their demonstrable capacity to deliver better canes.

Bonâme produced few useful canes. In 1907, after a decade of breeding and distributing cane cuttings to the estates, he acknowledged that both in the laboratory and in the field, none of the station's canes performed better than the imported Tannas.[51] He also worried that the Tannas might fail like all the previous varieties in widespread cultivation. In 1908, Bonâme went so far as to recommend that the Station Agronomique hire a botanist capable of supervising the breeding experiments.[52]

The science of sugar cane breeding was evidently still in its infancy during the years of the Station Agronomique's existence. Even so, breeders in Java, Australia and the British West Indies were obtaining better results.[53] The Chamber coveted other cane-growing regions' varieties and grew dissatisfied with the Station Agronomique's performance. Even Perromat's seedling canes were proving more useful than Bonâme's, and Perromat was a Mauritian nurseryman lacking in scientific credentials. One member of the Chamber reported to his colleagues the high cane yields in Hawaii, not realising that the islands had a longer growing season than Mauritius, and that they were using Lahaina cane, the Hawaiian name for Otaheite (Bourbon), a cane that had failed in Mauritius decades earlier.[54] Another sang the praises of the Uba cane in Natal, which he claimed could give twenty-five ratoons, forgetting that this cane contained a great deal of fibre.[55] Others clamoured for the introduction of the Badilla cane which had proved highly successful in Queensland.[56] Nobody needed reminding of how breeding had increased the sucrose content of the sugar beet, or of the remarkable sugar cane experiments in Java and Barbados.

The sugar industry turned once again to importing new varieties, conceding that the station's breeding programme might not produce a useful result for a long time. In 1900, the Chamber acquired seedling canes directly from the Imperial Department of Agriculture of the West Indies. In 1909, Bonâme obtained the Badilla cane and other varieties from the Queensland Department of Agriculture. Between 1904 and 1906, the Chamber even hired a local merchant firm to import new varieties from Barbados and British Guiana. At least one member made independent efforts to find canes abroad, while another may have been raising his own seedlings.[57] Over the course of the next two decades, some of the canes imported from British Guiana were cultivated on a small scale, and were also used as parent canes in breeding. During the decade of the 1900s, none of the newly imported varieties could match the Tannas, according to Bonâme's laboratory analyses.

Importing new cane varieties remained risky because of the possibility of introducing diseases and pests. Some members objected to the Chamber's new cane importations and proposed strict quarantine measures.[58] The government also feared diseases and required the Director of Forests and Gardens to inspect any imported canes, reducing the Chamber's autonomy to some extent. The Chamber still had an arrangement with the Pamplemousses Gardens to plant out the imported canes, and the sugar industry soon paid a heavy price for the negligence of government employees: plant inspectors failed to detect the introduction of a significant new pest, a beetle known as both *Phytalus smithii* and *Clemora smithii*.[59]

To make matters worse, some members of the Chamber raised questions about the Station Agronomique's utility. As early as 1901, the president wondered why the station was only spending Rs.20,000 each year, when the government was raising annual revenues of Rs.30,000 from the special export tax. Some members said the station could use the extra funds to make itself more useful, while others argued that the government ought to give the Chamber the entire Rs.30,000. One member said that if the station was doing no good, then letting the Chamber do the research directly would save the trouble of running an extra institution. Henri Leclézio, still the chairman of the Station Agronomique Committee, defended the station by accusing his sugar industry colleagues of making unsubstantiated claims. He said that his colleagues in the Chamber never bothered to read Bonâme's reports.[60]

Despite the efforts of Leclézio and his supporters, the station's detractors got the upper hand. Leclézio may have been right to accuse the Chamber of not taking full advantage of the station, but the impatient members persisted in their efforts to undo the institution so soon after its creation. One member argued that Bonâme's chemical analyses were a waste of everyone's time. Any estate could hire a chemist; what the industry needed was more general research on sugar cane. Even Virgile Naz, one of the station's founders, called the institution a *'corps inert'*, claiming that planters had few tangible results to show for their investment in the institution.[61] Between 1901 and 1906, members proposed using the station's extra money to fund an agent of the Chamber in India, to create a Station Bactériologique, to study how to achieve better telegraphy for the island, and to begin an information service within the Chamber itself. Between 1905 and 1910, the station's committee used the surplus funds to hire a veterinarian. One member even took it upon himself to travel to France to interview a botanist for a position at the station. He argued that although a good botanist would not come cheaply, the industry could ill afford falling behind in cane breeding.[62]

Members of the Chamber criticised Bonâme's experiments heavily. He produced little in the way of practical results, but he might have been able to buy himself more time if he had created a network of supporters. Had he worked harder to enlist the support of the Chamber, he might have become more sensitive to their desires and been able to manipulate them more carefully. The sociologist Bruno Latour gives a number of examples of scientists who have persuaded patrons to shift the ways in which they support research. He coins the term 'translation' to show how scientists and patrons 'enrol' each other into previously separate projects as a way to gain support for them. Of course, the price of this mutual enrolment is that scientists and patrons gain leverage over each other. Bonâme did not have the social or rhetorical skills that would have helped him to enrol the support of the Chamber. Or perhaps he worried that if the Chamber supervised the Station

Agronomique too closely, he would not be able to indulge his true love, chemistry.

In 1909, when a Royal Commission was investigating the Mauritian sugar industry, the Chamber asked for a state department of agriculture along the lines of the new, interdisciplinary department of agriculture in the British West Indies. Negotiations over the proposed department took several years, during which time it became clear that the Chamber wanted to exert a significant amount of influence over the department. In 1913, when the Mauritian department of agriculture was founded, the government did retain formal control, but the Chamber supervised the research and received most of the services, even though the department was supported by taxes on all Mauritian farmers.[63]

The Chamber's influence was especially notable in appointments to the department. It sought to appoint Mauritian laymen to research positions in the department, a proposition that appalled the Colonial Office. One Colonial Office bureaucrat described the proposed department entomologist as 'the ex-Curator of a slovenly and useless Museum', but he reserved his greatest contempt for the proposed department statistician, a journalist who was also the secretary of the Chamber: 'He is I believe intelligent – a Mauritian journalist has to be if he is to escape the Poor Law Department and the Prison Department.' The candidate's publications as the Chamber's secretary were decried as 'exactly the sort of thing which we don't want the Department of Agriculture to produce – long, tedious, and useless'.[64]

Even so, when the first director of agriculture arrived in 1913, he found that the Chamber of Agriculture had persuaded the governor to fill about half of the department's research positions with Mauritian laymen, including those two. The new director, Frank Stockdale, wrote to a friend that the Mauritians 'will have to dance to my tune'.[65] As things turned out, the Chamber continued to influence the appointments of British and Mauritian researchers to the department. Stockdale resisted the trend somewhat, but when he left Mauritius in 1916, the next director of agriculture, Harold Tempany, went with the flow and recruited Mauritians for the department, even training them in the new College of Agriculture according to metropolitan standards. This training helped teams of British and Mauritian botanists to make a number of important breakthroughs in cane breeding, principally because by the 1920s scientists were beginning to apply a rudimentary knowledge of plant genetics to the early hybridisation programmes. What is interesting, however, is that in the cane breeding programme, British breeders held supervisory positions and, judging by their publications, had a greater interest in theorising about cane genetics than their Mauritian assistants, who carried out the experiments. This may indicate that Mauritians were to play a subordinate role in cane breeding, but it may also indicate that the British recognised the skill of the Mauritians in manipulating experimental cane plants.[66]

This combination of British and Mauritian researchers helped to ensure that during the early twentieth century, Mauritian cane breeding became less empirical and more predictive. By comparison with other sugar colonies, Mauritian science might even be called advanced. However, it is also significant to note that all of the colony's Mauritian scientists were either Franco-Mauritians or assimilated Creoles of partly African descent; none of the researchers was an Indo-Mauritian, even though Indo-Mauritians made up more than half the population. It also never occurred to the Chamber or the government that the burgeoning classes of Indian peasants (who had bought land after their indentures expired) might desire to contribute to ongoing debates about funding and supervising sugar cane research institutions. Indian immigrants owned one-third of the lands under cane, but they were not yet fully acculturated to Mauritius, and they did not yet have the power to supervise botanical research. In Mauritius, lay knowledge of planters might influence the government's research on canes, and laymen from the plantations might even be trained according to imperial standards, but Indo-Mauritian peasants had only indirect access to the work of research institutions.

Peasants relied on plantation owners to get new canes, and they had no representation in the government. It was only during the depression of the 1920s and 1930s that peasants revealed that they, too, had the knowledge to produce and distribute new kinds of sugar canes. Peasant cultivators selected and distributed their own hybrid canes, known locally as Uba Marot. These were the result of a cross between a common variety of sugar cane cultivated on the plantations and a kind of reed that Hindus cultivated near their temples. This hybrid cane grew vigorously, which was a wonderful thing for peasants who were paid by the factories for the weight of the canes.[67] Alas, the Uba Marot canes also contained very little sucrose, which displeased the factories. One manager argued that 'Uba Marot should be treated as a poison by the factory'.[68] Other managers and owners agreed, so that eventually in 1937 the factories refused to accept any more Uba Marot canes.

As a consequence, many Indo-Mauritian peasants rioted and nearly succeeded in shutting down the entire Mauritian sugar industry. A commission of enquiry attributed the riots in part to the way in which the Department of Agriculture ignored peasant farmers:

The Department of Agriculture is the one directly concerned with questions of planting, but the Department does not appear to have taken any action to find a substitute for Uba until complaints began to come in about the cut of fifteen per cent. It is clear, however, that the Department of Agriculture will be largely concerned with any solution of the Uba question. We cannot but regret, however, that this question was not taken in hand at an earlier date. The Royal Commission of 1909 made an important recommendation to the effect that the Department of Agriculture should maintain the closest ties with the estates and the small planters. That recommendation has not been translated into action so far as we are aware, and we ourselves make a recommendation to this effect. Had the Department of Agriculture been closely in touch with the estates prior to the unrest, the problem created by the Uba cane might conceivably have been avoided.[69]

As it happened, the government's cane breeders, including senior British scientists and their Mauritian assistants, had just succeeded in breeding the first really good hybrid cane suitable for almost all Mauritian ecosystems: M134/32. This hybrid out-yielded all other available canes in terms of tonnage and it also had a reasonably good sucrose content, along with resistance to the island's principal pathogens. The government distributed this cane immediately to the peasant farmers, as well as to the estates, and it became the most widely grown cane for the next two decades.

This was a time when Mauritian politics were becoming increasingly democratic. The 1938 commission of enquiry into the Uba riots persuaded the governor, Bede Clifford, to legalise trade unions, to create a new labour department, and to establish a board to mediate disputes between planters and peasants. In 1947, another governor, Donald MacKenzie-Kennedy imposed a new constitution that extended the franchise to all Mauritian men and women over the age of twenty-one who were capable of writing a sentence in any language. The new Legislative Council was made up of nineteen elected and fifteen nominated members, three of whom were to be government officials, while the governor retained veto powers. In 1958, the British government ended the system of appointing members to the Legislative Council, and also ended the literacy requirement. Voters could now elect all of the Legislative Council, while the governor retained the power to appoint the most popular 'best losers' to the legislature, thereby ensuring the representation of all communities. In 1968, Mauritius became independent, and in 1992 it became a republic.

It was not a coincidence that peasants received wider access to new hybrid sugar canes during the transition to a new state. As Shapin and Schaffer argue in their study of science in seventeenth-century

England, 'Solutions to the problems of knowledge are solutions to the problems of social order.'[70] For most of the time between 1853 and 1937, the Royal Botanic Gardens at Pamplemousses, the Station Agronomique and the Department of Agriculture identified closely with the objectives of the colonial government and the Chamber of Agriculture. Producing new sugar cane varieties and hybrids became the key to producing sugar, tax revenues and a society organised around sugar production. In 1937, when the social order seemed on the verge of collapse, the Department of Agriculture reached out to the peasants and disseminated sugar canes directly to them.

The owners of sugar estates did resist the democratisation of agricultural research institutions. In doing so, they had some assistance from the colonial government, which was eager to provide economic stability even as the island was undergoing a political transformation. In 1953, the government allowed the Chamber to form a private research institution, the Mauritius Sugar Industry Research Institute (MSIRI). The government levied an export tax on sugar to support the institute, and also sold the Department of Agriculture's sugar cane research station to the new institute. The Department soon came under a parliamentary minister of agriculture, but it no longer had a branch that conducted research on sugar cane.

The MSIRI depended on the government for financial support. The government became more democratic over time, and so did the MSIRI's board of directors. This may have reflected the democratisation of the institution, or it may have reflected the success of the sugar industry in enrolling the support of erstwhile anticolonial and anticapitalist politicians. In any case, the MSIRI became one of the world's leading sugar cane research institutes, with numerous accomplishments in agronomy and breeding. Breeders produced so many new cane varieties that the problem of varietal decline was no longer significant. And during the late 1980s, the MSIRI cooperated with the government and the World Bank to establish semi-private Farmers' Service Centres (FSCs) for peasants. Among other things, the FSCs select the most reliable peasants and encourage them to cultivate new, experimental MSIRI hybrids, and to report to the breeders and to other peasants about the canes. This practice echoes the older practice, established in the 1860s, that allowed those laymen who held power to evaluate and supervise the work of scientific experts.[71] Along these lines, it is interesting to note that when I visited the FSCs in 1992, their largely male staffs were having difficulty reaching women farmers, who remain under-represented in both science and politics.

The history of cane science in Mauritius is significant because it contradicts some long-held assumptions about the relations between global capitalism, imperialism and plant research. In the most widely cited study of imperial agronomy, Lucile Brockway argued that British scientists in Kew set the research agendas in tropical colonies in order to exploit their resources.[72] Mauritius was dependent and its history is full of exploitation, but Mauritians from all classes and cultures were involved in debates about the production and distribution of cane varieties. In Mauritius, both science and power were constructed locally. This is not just to say that local politics became an external influence over scientific institutions, nor merely that scientific research supported the colonial state. It is to say that local power relations became internal to scientific methods and practices, and also that science became an integral part of political practice.

The rhetorical and social practices of sugar cane science were especially complex in colonial Mauritius, where cultural and linguistic differences as well as economic and political practices tended to keep people separate. Scientists persuaded farmers to take up new canes partly because the new canes yielded more sucrose, and also because the right kinds of lay people had supervised the research.

The case of Mauritius can help us to think more carefully about the way states define citizenship by making reference to science and technology: How do people begin to talk about access to particular kinds of science and technology as they are losing or gaining political power? To what extent do debates about science and technology clarify, obscure, or limit the political rights of subjects and citizens? The case of science and power in colonial Mauritius highlights all these concerns. It also raises questions about how debates about science and politics became embedded in research institutions, even as these institutions and the colonial state were changing in significant ways.

Notes

1. This article contains material published previously in my book, *Science and Power in Colonial Mauritius* (Rochester and London, 1997). I wish to thank the publishers for making it possible for me to recast some of those materials here.
2. C. Hamilton, *Terrific Majesty: The powers of Shaka Zulu and the limits of historical invention* (Cambridge, Mass., 1998).
3. S. Dubow, *Scientific Racism in Modern South Africa* (Cambridge, 1995).
4. R. B. Allen, *Slaves, Freedmen and Indentured Labourers in Colonial Mauritius* (Cambridge, 1999). In Mauritius, 'peasants' are usually called 'small planters', but the terms are essentially interchangeable, and 'peasant' seems more useful for comparisons.
5. A. S. Simmons, *Modern Mauritius: The politics of decolonisation* (Bloomington, 1982).
6. Storey, *Science and Power*, pp. 49–61.
7. Mauritius Chamber of Agriculture (MCOA), Minutes of 29 October 1868, pp. 50–1.

8 MCOA Minutes of 1 December 1868, p. 58.
9 MCOA Minutes of 24 October 1901, p. 18.
10 Horne to Dyer, 16 March 1881, Correspondence File (Corr.) 188.400, Royal Botanic Gardens, Kew (RBGK).
11 M. Biagioli, *Galileo, Courtier: The practice of science in the culture of absolutism* (Chicago, 1993); P. Dear, 'Totius in verba: Rhetoric and authority in the early Royal Society', *Isis*, 76 (1985), 145–61; S. Shapin and S. Schaffer, *Leviathan and the Air-Pump: Hobbes, Boyle, and the experimental life* (Princeton, 1985).
12 MCOA Minutes of 12 April 1877, p. 3.
13 MCOA Minutes of 17 May 1877, pp. 14–15; 22 August 1878, pp. 13-14.
14 G. C. Stevenson, *Genetics and Breeding of Sugar Cane* (London, 1965); Storey, *Science and Power*, pp. 74–6.
15 MCOA Minutes of 11 January 1871, pp. 109–10; P. de Sornay, 'Historique de la canne de graine à l'île Maurice', *Revue agricole et sucrière de l'île Maurice*, 10, no. 58 (1931), 125–6; G. Rouillard, *Historique de la canne à sucre a l'île Maurice, 1639-1989* (Port Louis, 1990), pp. 21-2.
16 MCOA Minutes of 3 May 1880, pp. 7-8.
17 D. Fitzgerald, *The Business of Breeding: Hybrid corn in Illinois, 1890-1940* (Ithaca, 1990), pp. 12–22; D. Headrick, *The Tentacles of Progress: Technology transfer in the age of imperialism, 1850-1940* (Oxford, 1988), pp. 215–16; M. W. Rossiter, *The Emergence of Agricultural Science: Justus Liebig and the Americans, 1840-1880* (New Haven, 1975).
18 B. Latour, *The Pasteurisation of France* (Cambridge, Mass., 1988). For a case study of Pasteur's methods in another colonial setting, see J. Todd, *Colonial Technology: Science and the transfer of innovation to Australia* (Cambridge, 1995).
19 Colony of Mauritius, *Papers Relating to the Sugar Crisis*, in the Mauritius Archives.
20 Summary of Newton's 'The Sugar Crisis'; copy of Pope Hennessy to Derby, 16 February 1885; Dyer to Herbert, 2 May 1885, all in RBGK *Mauritius. Agriculture. 1885–1900*, MR 11.4.
21 MCOA Minutes of 5 April 1887, pp. 17–21.
22 Storey, *Science and Power*, pp. 77–9.
23 Horne to Morris, 10 June 1890, RBGK Corr. 188.445.
24 Horne to Dyer, 9 December 1890, RBGK Corr. 188.451.
25 MCOA Minutes of 2 May 1888, Annex, pp. 1–2.
26 MCOA Minutes of 2 May 1888, Annex, p. 3.
27 MCOA Minutes of 24 September 1888, p. 18.
28 Colony of Mauritius, *Report on the Proposed Creation of a Station Agronomique* (Port Louis, 1890), in the Mauritius Archives.
29 MCOA Minutes of 12 May 1890, p. 10; 19 February 1891, p. 8; 18 June 1891, sheet enclosed between pp. 19–20; 14 December 1891, p. 33; 18 April 1898, p. 6; de Sornay, 'Historique de la canne de graine', p. 129; Rouillard, *Historique de la canne à sucre*, p. 22; H. Robert, *Sugarcane Varieties in Mauritius*, Department of Agriculture Bulletin No. 2, Statistical Series (Port Louis, 1915).
30 *Report on the Proposed Creation of a Station Agronomique*.
31 MCOA Minutes of 16 April 1891, Annex, pp. 1–2.
32 MCOA Minutes of 19 February 1891, pp. 7–8.
33 MCOA Minutes of 19 February 1891, pp. 7–8.
34 *Report on the Proposed Creation of a Station Agronomique*, p. 15 and *passim*.
35 *Report on the Proposed Creation of a Station Agronomique*, p. 15.
36 Shapin and Schaffer, *Leviathan and the Air-Pump*, p. 55.
37 *Report on the Proposed Creation of a Station Agronomique*, p. 15.
38 Lees to Knutsford, with minutes by Wingfield, 9 January 1891, PRO CO 167/661.
39 H. Leclézio, 'Minutes of Proceedings of the Committee at a Meeting Held at Government House, Port Louis, the 29th of June 1892', enclosed with governor's despatch to the Colonial Office of 12 July 1892, PRO CO 167/669.
40 'Director of Station Agronomique', 15–18 August 1892, PRO CO 167/669 no.15464.
41 'Rules for the Working of the Station Agronomique', RBGK *Mauritius. Agriculture.*

1885–1900, MR 11.4, p. 168; Minute by Wingfield, 4 November 1892, PRO CO 167/670.
42 Station Agronomique, *Rapports et bulletins*, 1900–11.
43 Station Agronomique, *Rapports et bulletins*, no. 18, pp. 14–15.
44 Rouillard, *Historique de la canne à sucre*, p. 22.
45 Rouillard, *Historique de la canne à sucre*, pp. 19–20.
46 Colony of Mauritius, *Annual Report of the Royal Botanic Gardens, Pamplemousses*, 1891.
47 MCOA Minutes of 12 October 1894, p. 25; 3 April 1895, p. 18; 15 July 1895, p. 20; 2 December 1895, p. 26; 15 January 1896, pp. 1–2; 7 April 1896, p. 3 and Annex, pp. 2–3; 17 November 1897, p. 30.
48 MCOA Minutes of 7 April 1896, p. 3.
49 MCOA Minutes of 28 December 1898, p. 62; 15 February 1900, p. 3; 24 October 1901, p. 18.
50 Station Agronomique, *Rapports et bulletins*, Annual Reports of 1900, pp. 20–1; 1902, pp. 24–6; 1904, pp. 17–19; 1908, pp. 49–51.
51 Station Agronomique, *Rapports et bulletins*, no. 15, pp. 1–2.
52 Station Agronomique, *Rapports et bulletins*, no. 18, 1908, pp. 14–15; 1909, pp. 25–7.
53 Storey, *Science and Power*, pp. 87–8.
54 MCOA Minutes of 11 April 1901, pp. 8–9.
55 MCOA Minutes of 7 April 1904, p. 20.
56 MCOA Minutes of 29 July 1908, pp. 2–3.
57 MCOA Minutes of 26 October 1900, p. 20; 3 November 1904, pp. 24–5; 29 December 1905, p. 27; 7 March 1906, Annex; 25 April 1906, Annex; 7 February 1907; 8 November 1907, Annex B; 30 December 1908; Station Agronomique, *Rapports et bulletins*, Annual Report of 1909.
58 MCOA Minutes of 7 April 1904, p. 19; 18 August 1904, p. 2; 3 November 1904, pp. 24–5.
59 MCOA Minutes of 3 August 1911.
60 MCOA Minutes of 11 April 1901, pp. 8–9.
61 MCOA Minutes of 3 July 1901, pp. 12–13.
62 MCOA Minutes of 17 October 1902, p. 34; 27 October 1902, p. 36; 27 April 1904, Committee Report Annex; 18 August 1904, p. 2; 21 June 1905, p. 16; 21 June 1905, p. 17; 16 August 1905, p. 21; 15 November 1905, pp. 25–6; 2 May 1906; 18 July 1906, pp. 11–12; 8 August 1906, p. 16; 16 August 1906, p. 19; 1 July 1910, p. 2.
63 Storey, *Science and Power*, pp. 103–8.
64 Harding to Collins, 25 April 1912, PRO CO 167/801.
65 Stockdale to Hill, 7 June 1913, RBGK Corr. 188.554.
66 Storey, *Science and Power*, pp. 110–12, 115–17, 139–41.
67 Colony of Mauritius, Department of Agriculture, *Annual Reports*, 1927, p. 16; 1928, p. 9; 1933, pp. 16–17; Sugarcane Research Station, *Annual Report*, 1933, p. 35; G. C. Stevenson, *An Investigation into the Origin of the Sugarcane Variety Uba Marot*, Sugarcane Research Station Bulletin No. 17 (Port Louis, 1940), pp. 4–7; C. A. Hooper *et al.*, *Colony of Mauritius. Report of the Commission of Enquiry into Unrest on Sugar Estates in Mauritius, 1937* (Port Louis, 1938), pp. 133–5.
68 F. North-Coombes, *Mes champs et mon moulin* (Port Louis, 1950), pp. 355–6.
69 Hooper, *Report of 1937*, p. 143.
70 Shapin and Schaffer, *Leviathan and the Air-Pump*, p. 332.
71 Storey, *Science and Power*, ch. 6.
72 L. H. Brockway, *Science and Colonial Expansion: The role of the British Royal Botanic Gardens* (New York, 1979).

CHAPTER THREE

A commonwealth of science: the British Association in South Africa, 1905 and 1929[1]

Saul Dubow

Setting

On 15 August 1905 a party of some 200 official members of the British Association for the Advancement of Science arrived in Cape Town on board the Union Castle Liner, *Saxon*. The voyage had been pleasantly uneventful and the visitors occupied their time with an extensive programme of lectures and discussions, games and entertainments, and scientific experiments besides.[2] The boundaries between amusement and scientific investigation were not rigidly maintained. It was in the jovial spirit of scientific adventure, we must therefore assume, that the head of the president of the Association, the mathematician and astronomer George Darwin (son of Charles) consented to having his head measured with a pair of calipers as he reclined on deck in a comfortable cane armchair.[3]

The ostensible purpose of the visit of the British Association was to hold its annual meeting together with the newly formed South African Association for the Advancement of Science, members of which had been enrolled as associates of the British Association by special arrangement. But there were important ideological issues involved too and these constitute the focus of this chapter. In looking closely at the visits of the British Association to South Africa, first in 1905, and again in 1929, I wish to address some of the complex relationships between science, imperialism and colonial nationalism, and to touch as well on the place of science in constructions of racial and ethnic identity. These issues, I argue, were of central concern to the organisers of the British and South African Associations. They raise questions about science, conceived of as a universalistic, transcendent and objective set of practices, and its use as a means of promoting particularist and partial political purposes in the period of transition from empire to commonwealth.

A COMMONWEALTH OF SCIENCE

A precedent for overseas meetings of the British Association had been set with its visits to Montreal in 1884 and Toronto in 1897. However, whereas previous gatherings took place in a single city, the 1905 visit involved travelling through the country and was therefore thought of – in a phrase redolent with meaning – as a 'South African meeting'.[4] A total subvention of £6,000 was provided by the colonies of the Cape, Transvaal, Orange Free State and Natal towards the sea passages of most of the overseas visitors, whose numbers totalled some 380. The head of the Union Castle Mail Steamship Company, Sir Donald Currie, offered reduced fares for the Atlantic crossing, and pledged his personal support for the enterprise on the grounds that it was 'thoroughly in accord with the spirit of Imperialism'. A private South Africa Fund was subscribed in Britain to cover additional expenses, and free railway passes were granted to cover the extensive internal travel arrangements.[5]

The expedition was a major undertaking, with the official proceedings held over the course of two weeks in both Cape Town and Johannesburg. Visits were also arranged to centres such as Durban, Pietermaritzburg, Bloemfontein and Kimberley, at which lectures were delivered and civic receptions laid on. Optional side tours were arranged to Victoria Falls, Bulawayo and Beira. Elaborate arrangements for the visit were made by the specially constituted South African committee of the British Association in London which had the responsibility of directing overall planning. The burden of much of the preparatory work was borne by the Cape Town-based committee, consisting of the Royal Astronomer and founding president of the South African Association, Sir David Gill, and the marine biologist John D. F. Gilchrist. To ensure high-powered political support, several vice-presidents of the British Association were appointed, including the High Commissioner for South Africa, Lord Selborne, and his predecessor, Lord Milner; the Governors and Lieutenant-Governors of the four colonies of South Africa; the Administrator of Southern Rhodesia; and the mayors of cities included in the Association's itinerary.[6]

The visit of the British Association was marked by lavish hospitality on the part of private individuals and the royal progress of the scientists was punctuated by a succession of enthusiastic civic welcomes. Towns competed with each other in fussing over their illustrious guests and bid to be included in the offical programme. At every stop reception committees were busily active, local excursions were arranged, and a profusion of toasts, votes of thanks and enthusiastic tributes were traded between local and visiting dignitaries. Great pride was taken in the precision of the organisation and its planning. Every visitor was allotted a particular seat on one of the four special trains; on arrival at any of the towns at which they were staying, hosts were able to assemble on the platform in the order in which their guests would alight.[7]

In Johannesburg, the visit was marked by one of the greatest social events ever held in the city: a mayoral party given at the Wanderers' Hall attended by a 'brilliant crowd' of some 1,800 notables.[8] One of the most remarked upon outings was a visit to the Mount Edgecombe sugar mill outside Durban where the scientists delighted in photographing a dance performed by hundreds of 'Zulus' in the presence of the Governor of Natal. The railways as well as local publicity organisations printed lavish programmes with illustrated essays highlighting the special features of particular regions and cities.[9] Sightseeing was evidently as much a priority as serious scientific discussion, a fact attested to by the rather low attendance rates at sectional meetings.[10] Indeed, one visitor extended his warmest gratitude to his South African 'kinsmen', drily thanking those 'who helped to make our trip to their continent such a delightful and enjoyable picnic'.[11]

If elements within the British Association were concerned by the apparent lack of serious scientific purpose in the visit, their fears were outweighed by the enthusiasm and publicity that was generated. The objects of the British Assocation were always wider than purely academic and, since its foundation in 1831, the organisation had proved adept in reinventing itself to suit new realities and constituencies. Originally conceived in the age of parliamentary reform as a bridge between specialists, practitioners and the general public, the British Association proved responsive to shifts in the political climate. In the early 1880s it had been beset by a sense of malaise and decline, reflected in falling attendances at yearly meetings. One way in which the Association reinvigorated itself was by reaffirming links with provincial and amateur scientific societies. A parallel initiative, graphically symbolised by the decision to meet in Canada in 1884, was the decision to cultivate links with the empire. This reorientation was given particular impetus by the Association's growing realisation that it was losing its way as scientific endeavour in Britain became increasingly specialised and professionalised. Involvement in the empire therefore offered new possibilities for the British Association to reassert its prestige and to fulfil its historic role as a populariser of scientific method and culture.[12]

Michael Worboys places particular emphasis on the British Association's growing concern with empire in terms of the economic and political imperatives of late nineteenth-century social imperialism, a doctrine which sought to contain domestic tensions and crises both by projecting these outwards into the empire and by drawing the colonies into closer economic contact with Britain. The Association's self-proclaimed 'imperial mission' was an ideological and cultural expression of these processes. Thus, Worboys indicates how the Association, in identifying itself with the 'universality of science' and ideas of progress, served as a metaphorical embodiment of the idea of empire.[13] In the case of Canada, as Gale has shown, the 1884 meeting of the British Association marked 'a dramatic demonstration of the important role which science could play in peacefully federating the interests of the nation states of the empire', while at the same time 'boosting and reaffirming the Association's role as an organizing centre of research within the provinces and the Empire at large'.[14]

A similar point can be made in the case of South Africa. The 1905 visit was intended to confer status on the newly constituted South African Association for the Advancement of Science (S2A3) whose immediate history went back to a meeting held in Cape Town in July 1901 at which a proposal was made by Theodore Reunert to establish an annual congress of engineers. The term 'Engineer' was defined in the broadest of terms so as to encompass 'not only those engaged in the utilization of Science, but those also whose lives and interests are occupied in the pursuit of Science for its own sake'.[15] The idea of forming an association similar to the British Association was canvassed at the gathering and, at a further meeting, it was resolved to constitute the S2A3 on this model.

In 1900 Sir David Gill attended a meeting of the Council of the British Association which had expressed interest in holding one of its annual meetings in South Africa. As a fellow of the Royal Society, as joint president of the S2A3 and the South African Philosophical Society, and with his recent knighthood, his prestigious position as royal astronomer and his extensive connections, Gill was well placed to bring this about. From 1902 he became actively engaged in gathering support for the visit from the mayors and corporations of Cape Town and Johannesburg, the Chambers of Commerce of these cities, the stock exchange and the Chamber of Mines. The end of the Anglo-South African War now made such a visit possible. The prime minister of the Cape, Gordon Sprigg, was therefore approached with a view to bringing out the British Association in 1905. He agreed and promised to provide material assistance. Gill was also successful in securing the support of Joseph Chamberlain who visited South Africa at the beginning of 1903. He informed the colonial secretary that the British Association would be going 6,000 miles to help the 'work of conciliation in South Africa'.[16] In his address to the S2A3 that year, Gill underlined the role that science could play in healing the bitterness of war and in cementing the imperial connection. 'Science', he argued, 'knows no nationality, and forms a meeting-ground on which men of every race are brethren, working together for a common end – and that end is truth.'[17]

The 1905 meeting

The British Association arrived in 1905 to a traumatised country poised uneasily between the experience of high-handed imperial domination and the beginnings of self-government and political renewal. The recent departure from South Africa of Milner as High Commissioner, the formation of the Het Volk Party and the publication of the Lyttelton Constitution providing for self-government in the Transvaal, pointed to a measure of rejuvenation in the political process. There were also welcome signs of economic recovery after the harsh commercial depression that followed the end of war. On the other hand, Campbell-Bannerman's victory in the British general election – and the decisive effect that this would have on the domestic political situation in South Africa – lay some months in the future. The British Association's visit therefore occurred during a delicate political hiatus. As the assistant secretary of the British Association delicately put it in an interview with the *Morning Post*: 'The South African colonies are passing through rather a trying period in their history. It is a matter of common knowledge that times are none of the best.'[18]

This sense of unease was reflected in the cautious approach taken by the British Association. A recurring theme in the many speeches delivered at public functions celebrating its visit was the healing role that science could play, but this sentiment was principally directed to 'British' South Africans with gestures of reconciliation directed only obliquely to 'Boers'. The tone was set by the Governor of the Cape, Sir Walter Hely-Hutchinson, when he welcomed the British Association to Cape Town at the start of its visit. Noting that the occasion was 'of no ordinary importance, whether in the history of scientific inquiry or in the history of the relations of the United Kingdom with the British dominions beyond the seas', he expressed hope that an 'important step had been taken in drawing closer together the bonds of the brotherhood of science' and, in so doing, 'promoting and developing brotherly feeling between his Majesty's subjects in South Africa and the motherland'.[19]

The idea that science had no political boundaries undoubtedly sent out a message of inclusiveness. But, for the most part, this sense of unity was cast in the context of a shared sense of imperial belonging. In Johannesburg, the Lieutenant-Governor of the Transvaal, Sir Arthur

Lawley, expressed regret that South Africa was a land encircled by walls constructed from 'prejudices of caste, colour, and race' born of past feuds and divergent interests. And he suggested that, like Jericho, such walls 'were only to be levelled by trumpet blasts of knowledge and of the wisdom which knowledge bore in its train'.[20] Likewise, Sir Richard Jebb hoped that, in promoting mutual understanding, the visit of the British Association would not only strengthen imperial ties, but would also facilitate more cordial relations between the 'two white races in the continent'.[21] Others were more inclined to jingoistic triumphalism. Thus, Col. Sir Colin Scott Moncrieff spoke 'as a Briton to Britons' even as he noted that 'science was of no nation'. By this he meant that the Association's visitors included scientists from America, Europe and even far-off Japan. His expression of 'kindly sympathy with the brave people who were but lately their enemies, but now the subjects of the King', can only be interpreted as a form of smug condescension.[22]

In the many tributes to the Association's role in reaffirming 'our English-speaking brotherhood', the 'strengthening of ties between the South African colonies and the Mother Country', and the formation of more links in the 'chain of civilization',[23] there is a distinct sense that the Association was determined to cement imperial unity and to reiterate the fact of British supremacy. The extensive programme of travel satisfied the needs of tourism, but it was also designed to encompass the imperial domain. Care was therefore taken not to marginalise areas of the country which, though visited, remained beyond the official scientific programme, and the existence of inter-colonial rivalry was recognised in placatory fashion.[24] Inevitably, choices had to be made and some towns were left out of the itinerary. Despite its historic English character and its strong claims to be an educational centre, the invitation from Grahamstown was declined. So, too, was an opportunity to visit the Lovedale Institution where Gill promised 'they would see more than 1,000 natives of the different races and obtain from Dr. Stewart (who is probably the most interesting living authority on the subject) all his views with regard to their capacity, the possibilities of their future, etc.'[25]

Some of the most poignant and affecting excursions were those arranged to the battlefields of the recent war. At Colenso, the visitors arrived in luxury trains and were able to spend time finding cartridges, unspent bullets and shell fragments. James Stark Browne noted approvingly that the war graves were being maintained in beautiful condition by the Guild of Loyal Women.[26] At Paardeberg, the scene of General Cronjé's surrender, skeletons of horses, meat and biscuit tins and other raw war debris were abundantly evident. Both the *Star* and

The Times reported that the visitors 'were generally delighted with their experience'. As if to underline the sense of victory and the restoration of political order, a party of intrepid scientists trekked across the veld to Kimberley; they were guided by a farmer named 'Joe' who was said to have fought with Cronjé but now gamely entertained them by singing Boer songs around the fire.[27] The *Diamond Fields Advertiser* echoed this theme: whereas 'the Fates' had previously decreed that 'the only missions to achieve any large measure of success in our direction should partake of a somewhat warlike nature', the new mission came 'in the peaceful and cosmopolitan garb of science'; it was to be hoped that their 'patriotic aspirations' would ultimately be realised.[28]

A distinct note of imperial triumphalism was sounded when the Association visited the Victoria Falls in Southern Rhodesia. The president of the Association, Professor George Darwin, was given the honour of opening the new railway bridge over the Zambezi for passenger travel. Referring to the occasion as the 'crowning glory of the tour', Darwin took the opportunity to reflect on the achievements of Cecil Rhodes and the technological marvels of modern railway travel. He found himself unable to resist quoting from his great-grandfather, Erasmus Darwin, who had anticipated the potential wonders of steam, and he proceeded to muse about how his illustrious ancestor would react if he could see his great-grandson 'declaring a railway bridge open in the heart of Equatorial Africa'. A telegram from the British South Africa Company underlined the unmistakeable symbolism with the words: 'Very fitting that foremost representative of science should be associated with inauguration of modern engineering. Regret founder of country is not alive to witness realization of part of his great ideal.'[29]

The 1905 meeting was widely considered to have been a great success in all respects; thus there were many tributes to the contribution made by the British Association in helping to advance the development of scientific work in South Africa, as well as to the generosity of the South African hosts. The papers delivered by the Association's visitors covered a typically wide range of subjects and the characteristic balance between pure and applied science was consciously maintained. Papers with a particularly South African focus were prepared for publication in four special volumes (with the co-operation of the South African Association) though their quality gave rise to concern.[30] A few even made it into the general *Report of the British Association*.

Reflecting the pre-eminence and leading organisational role of the South African Association's foremost scientist, Sir David Gill, there was a large concentration of papers on astronomy and geodesy. The tradition of astronomy in South Africa was, indeed, one of the foremost and earliest examples of science in South Africa (if not always science of South Africa). One of the concrete results of the Association's visit was a renewed commitment to complete the geodetic survey of Africa from south to north which had been inaugurated by Lacaille during his time at the Cape in the 1750s. This programme was one of a number of ambitious survey projects initiated by Gill from the 1880s onwards. It had bearings on theoretical questions relating to the size and shape of the earth and its distance from the heavens, as well as having vital practical implications for the processes of triangulation and mapping of colonial boundaries.[31] South African related material was also prominent in Sections C (Geology) and E (Geography) owing in part to the country's rich array of mineral deposits and the development of the diamond and gold mining industries from the 1860s. Nineteenth-century pioneers of South African geology such as Andrew Geddes Bain, Andrew Wyley, P. C. Sutherland, G. W. Stow and E. J. Dunn were well known to the British Association visitors. The timely publication in 1905 of the first comprehensive textbook on the geology of South Africa was indicative of the relative maturity of scientific work in this field.[32]

A considerable boost to anthropological and ethnological work was provided by the presence of A. C. Haddon who attempted to survey the existing field and to map out directions of further research.[33] Yet the disparate nature of the contributions to Section H (Anthropology) indicates how undeveloped South African anthropology was at this time – notwithstanding contributions by scholars such as L. Péringuey on the 'Stone Age in South Africa', Henry Balfour on 'The musical instruments of South Africa' and H. A. Junod on 'The Thonga tribe'. Perhaps the most significant – and certainly the most controversial – paper in this section was David Randall-MacIver's report on Great Zimbabwe which concluded that 'the Rhodesian ruins were the work of a native race closely akin to those at present inhabiting the country'. Great Zimbabwe, MacIver maintained, was all the more interesting as a product of the southern African region rather than as a 'parasitic growth from Arabia' – even if this conclusion destroyed the romance of its supposed links with biblical antiquity.[34]

From a political point of view the most notable contribution was in Section F (Economic Science and Statistics) where Howard Pim delivered an address entitled 'Some Aspects of the Native Question'. The importance of this paper in defining and outlining the concept of racial segregation in South Africa is now well established and it attracted considerable attention in the press at the time.[35] So, too, did Sir

Richard Jebb's presidential address to Section L (Educational Science) on 'University Education and National Life'. In the course of his discussion of the writings of Newman, Huxley, Arnold and Sidgwick, Jebb insisted on the interdependence of scientific and literary culture. Jebb's remarks on the need for a teaching university in South Africa and his positive comments on the achievements of the new civic universities (especially Birmingham) in satisfying the demand for professional, technical and scientific as well as general education, were widely noticed – particularly in Johannesburg.[36]

Despite the importance of the British Association's 1905 visit in drawing attention to, and providing a focus for, intellectual endeavour in South Africa, systemic weaknesses were all too evident. As Jebb's remarks made clear, there was as yet no fully fledged teaching university in the country, there were few professional bodies and organisations in existence, resources were scarce and there was little if any overall coordination of research. The indebtedness of South Africa to the British Association was therefore all too evident and the latter did little to downplay its elevated status. Although kindly remarks were made about the progress of South African science these were mostly delivered in the form of gestural encouragement.[37] In short, the British Association was on display and the gratitude of the locals sometimes came close to obsequiousness.[38]

Institutions and nationhood

In the absence of large-scale institutions dedicated to the production and dissemination of basic intellectual and scientific knowledge, the S2A3 and the Philosophical Society were important bodies. They were in turn supported by a number of professional organisations which helped to foster the development of intellectual communities and scientific awareness. Many of these were established during the final decades of the nineteenth century, including the Institute of Mechanical Engineers and the Association of Engineers and Architects (1892), the Chemical and Metallurgical Society of South Africa (1894), the Geological Society of South Africa (1895), the South African Society of Electrical Engineers and the Transvaal Medical Society (1897).[39] The medical profession was an especially important focus of scientific activity. Doctors were often amateur scientific investigators of note and, importantly, they were based in the country towns and hamlets as well as the cities. Publications like the *South African Medical Journal* and the *South African Medical Record* existed principally to promote the collective interests of doctors but they also functioned as a vital outlet for the discussion and dissemination of medical and associated scientific research. A similar mix was evident in the *Cape Law Journal* which played a vital role in regulating the legal profession, circulating law reports and contributing to a national tradition of jurisprudence.

Another significant focus of scientific activity was the cluster of museums which grew up in the second half of the nineteenth century. The oldest and most prestigious was the South African Museum in Cape Town which traced its origins back to 1825 but which was re-established almost from new in 1855. The Albany Museum in Grahamstown was founded in the same year. In Natal, museums were established at Durban in 1877 and Pietermaritzburg in 1904. The Boer republics boasted similar institutions: a National Museum of the Orange Free State had been created in 1877, while the South African Republic created its Staats Museum in 1892. These museums were principally civic institutions but they could also serve as a focus for broader national aspirations. The initiative for the Staats Museum, for example, had come from W. J. Leyds, a Dutch-born cultural entrepreneur and close associate of Kruger. Although developed as a general museum with a strong emphasis on natural history, it was specifically intended to promote the republican cause by stressing Afrikaner cultural history and it attracted considerable popular interest as a result.[40] Competition between the South African and the regional museums was often keen. One index of such rivalry, together with an increasing concern with professional standards of museology and research, can be seen in the development of house-journals, beginning with the *Annals of the South African Museum* in 1898. By 1907 similar publications were being produced by the Albany, Natal government and Transvaal museums.

In the context of regional rivalries the newly formed S2A3 played a significant role helping to sew the sinews of a wider national scientific identity. Amongst its declared constitutional aims was 'to promote the intercourse of Societies and individuals interested in Science in different parts of South Africa' and 'to obtain a more general attention to the objects of pure and applied Science'.[41] From the outset, therefore, the S2A3 sought to include not only those with a professional or academic interest in science, but also amateurs and enthusiasts.[42] Its catholic approach was reflected in its impressive growth in membership figures: 268 'foundation members' in 1902; well over 700 members and associates at the end of the first meeting in 1903; and 1,322 in 1906[43] – a figure never surpassed in the twentieth century. Theodore Reunert recalled that it was founded as an 'act of faith' 'at a time of great and universal despondency'.[44] Its leaders saw their mission, at least in part, in terms of drawing South Africa together and in assuaging the bitterness of war. This emerged as a major theme in David Gill's 1903 presidential address wherein he argued that periods of political turmoil were almost invariably followed by 'intellectual progress and development' and that the Association could help to facilitate this process.[45] This politically inclusive but avowedly imperial-centric view of science was reiterated in 1907 by the Grahamstown-based journal, *African Monthly*, when the four South African colonies were beginning to explore actively the possibilities of unification. At a time when symbols of national unity were being keenly sought after, it noted the importance of a unitary parliament if the 'ideal of a South African nation, British in character, and therefore Imperial in tendency' were to prevail. Equally important in breaking down 'petty localism' and establishing 'true' nationality was the common desire on the part of individuals to cooperate with one another; the creation of S2A3, it noted, was welcome evidence of this process.[46]

Although it is not clear whether the *African Monthly* envisioned the S2A3 as a metaphorical South African parliament, had it done so it would only have been echoing a recurrent image of the British Association as the 'parliament of science'. That view reflected the British Association's desire to serve a range of scientific constituencies and interests and underlined the fact that, in acting as an advocate for science, it inevitably participated in the realm of politics; indeed, parallels were drawn between the foundation of the British Association in 1831 and the passage of Russell's first Reform Bill a year later.[47] The S2A3 may not have seen itself in quite the same terms. But it did play a significant role in the politics of symbolic unity, particularly as far as anglophone South Africans were concerned. This was implicit in the very structure of the S2A3 which was made up of local committees and governed by a Council composed of members drawn from the major centres of the country. By the time of Union the S2A3 had met in all four colonies. It was thus one of only a few nationwide organisations (including, perhaps, the churches and the medical profession) enjoying local as well as international links.

The focus on scientific advancement was a conspicuous feature of the reconstruction period following the Anglo-South African war. It has often been noted that, whatever the specific failures of Milner's cultural and political efforts to anglicise South Africa (with respect to educational and immigration policies, for example), his campaign to reorder the social, economic and administrative structure of the country was both extensive and marked by a high degree of success. The impetus given to modernisation placed a premium on the role that science and technology should play in the creation of a new society and this had both ideological and practical dimensions. For Milner and his followers technical competence was not only proof of the transformative and progressive capacity of a reawakened British imperialism, it also had the advantage of being capable of being represented as the politically acceptable face of state intervention.

An important index of the post-war technocratic tendency of government was the encouragement of scientific methods in agriculture which, as Donald Denoon points out, was the one aspect of Milner's work for which Louis Botha was willing to praise the post-war administration.[48] The creation in the Transvaal of a department of agriculture in 1902 under the direction of F. B. Smith provided an important stimulus to the adoption of modern scientific methods which attracted the particular interest of ambitious young Afrikaner modernisers.[49] Immediately following the grant of self-government to the Transvaal in 1908 and on the invitation of Botha, now prime minister, Dr Arnold Theiler, a Swiss-born zoologist and bacteriologist, created a new and well-funded veterinary institute at Onderstepoort near Pretoria. After Union, Onderstepoort came to coordinate veterinary research work in South Africa, absorbing the existing laboratories in Grahamstown and Natal as satellites.[50] Its achievements in applied agricultural and veterinary research, together with the creation of a centralised Department of Agriculture in 1911, came to be celebrated in the modernising nationalist idiom of man's progressive technical mastery over nature and, by implication, of white South Africans' special scientific knowledge of, and control over, a hostile African environment.[51]

Aside from new departments of state such as the Agriculture department, a number of nationally significant scientific institutions and projects were inaugurated in the immediate post-Union era. Amongst the most important were the South African Institute for Medical Research (1912), jointly funded by Union government and the mining industry; the Geological Survey of South Africa (1910);[52] the first national statistical survey (1911), the Meteorological Office (1912); and the National Botanic Gardens in Kirstenbosch (1913). The creation of national scientific institutions required delicate political balancing acts because provincial rivalries were often aroused in the process of forging a wider sense of national identity. Regional institutions often competed for favour and funding and, inevitably, some lost out as a result of national consolidation. In the case of museums, for example, unification secured the premier position of the South African Museum in Cape Town which, together with the Transvaal Museum, now benefited from a Union government grant and were awarded premier status. Conversely, the new national dispensation worked to the detriment of institutions like the Albany Museum which now came under the aegis of the provinces and were treated as being of secondary importance.[53]

The establishment of Kirstenbosch is especially revealing of the complex interaction between local, national and imperial connections. Its immediate origins go back to the 1880s and 1890s when a renewed phase of locally oriented botanical activity was inaugurated by researchers such as Harry Bolus, Peter MacOwan and Rudolph Marloth at the Cape, John Medley Wood in Natal, and Selmar Schonland in the eastern Cape.[54] Writing in *The State* in 1909, a literary and political venture created by members of Milner's kindergarten to advance the political cause of closer union, Bolus couched his appeal to preserve the Cape's distinctive flora in terms of the awakening of 'a higher patriotism in South Africa'. He feared that Table Mountain was rapidly being denuded of heaths, orchids and anemones by flower sellers and casual pickers and, in calling for the creation of a special reserve in order to protect the country's native flora, he cited the example of the Yellowstone Park in the United States and the National Trust in England.[55]

This plea was reinforced by Harold Pearson who came from Kew Gardens in 1903 to take up the newly established Bolus chair of botany at the South African College. Along with Neville Pillans, Pearson was the leading figure in the move to establish the Kirstenbosch botanic gardens and was personally responsible for selecting its site. He couched his appeal for a state botanical garden partly in the familiar language of utilitarian objectives (commercial and economic possibilities, medical and agricultural research) but principally on the grounds that the pursuit of 'science for its own sake' was a mark of civilised society. Tying this ideal to the quest for national identity, he portrayed the natural world as uniquely expressive of the nation, arguing that there was 'nothing more truly "South African" than its indigenous flora'.[56] Thus:

> The South African Botanic Garden cannot be merely an economic undertaking; it must also be an expression of the intellectual and artistic aspirations of the New Nation whose duty it is to foster the study of the country which it occupies, to encourage a proper appreciation of the rare and beautiful with which Nature has so lavishly endowed it.[57]

The creation of Kirstenbosch as a National Botanic Garden (encouraged by the S2A3 and aided by a small grant from the Botha government) gave expression to this national sentiment and reflected the brief blaze of optimism which characterised the immediate post-Union period. From the beginning, the cultivation, display and study of the indigenous flora of South Africa was the raison d'être of the

Kirstenbosch. In this sense it was distinctive because, unlike botanic gardens elsewhere in the empire which were devoted to the collection and study of foreign flora and which were closely tied into the colonial network of which Kew formed the hub, Kirstenbosch was additionally intended 'to be a Garden for the study of the flora of South Africa itself – the country in which it was situated'.[58] The establishment of Kirstenbosch at the side of Table Mountain can therefore be seen as the culmination of a process whereby the flora of South Africa, which had previously attracted the enthusiastic attentions of botanists and collectors as 'marvels' or 'curiosities', now became truly indigenised and were seen as valuable in their own right.[59] As a garden featuring unique and local rather than exotic botanical specimens, Kirstenbosch proclaimed South African nationhood. At the same time it could also be seen to express a more restricted and specific sense of Cape-ness.[60]

Also bound up with the struggle for national and regional identity was the movement to create a system of tertiary education. At the time of the British Association's 1905 visit there was only one 'university', the University of the Cape of Good Hope, whose foundation in 1873 was one of the first tangible rewards for the grant of responsible government to the Cape. In reality it was little more than an examining body (modelled on the old University of London) designed to serve the disparate teaching colleges spread about the country. There were thus, in 1905, around 50 teaching staff and fewer than 600 students. By contrast, when the British Association next visited South Africa, in 1929, the university system consisted of around 500 lecturers (of whom nearly half were professors) and the residential student body had grown to more than 7,000.[61]

This rapid growth (albeit from a small base) was both conditional upon and indicative of the process of political unification. Propagandists of closer union in the first decade of the century made much of the need for a national policy on university education. For some, like Chief Justice J. H. De Villiers, the creation of a national teaching university was 'a necessary first step under Union'.[62] From 1908, when an inter-colonial conference was called in Cape Town in order to discuss the future of tertiary education, various government commissions considered what had become known as the 'University Question' – though with a conspicuous lack of success. In 1916 several key university acts were passed as a result of which the modern distribution of anglophone and Afrikaner white tertiary educational institutions was established. But the initiative to create a single national university proved a failure, partly because of the individual aspirations of the existing educational institutions, but also because of ongoing regional competition for cultural and political primacy between the Cape and the Transvaal and, in particular, because of growing sectarian conflict between English- and Afrikaans-speakers.[63] The 1928 Van der Horst Commission considered that the ideal of a single national university in South Africa had proved impractical and that the historical trend was in the opposite direction.[64] Eric Walker shrewdly commented at this time: 'Politically and economically, South Africa had just moved centripetally; academically, in the very act of political union, it had rushed confusedly in the opposite direction.'[65]

One of the most articulate and persuasive interwar champions of South African science and, especially, of its role in the creation of nationhood, was Jan Smuts. His understanding of this link was bound up in his personal holistic vision whereby the local and the particular was also part of the transcendent and the universal. Whereas many extreme forms of nationalism seek to root themselves in mystic notions of the soil, forest or landscape, Smuts's 'patriotism of place' and sense of national belonging was defined through the more detached perspective of the philosopher and naturalist.[66] As a noted botanist and philosopher Smuts considered himself qualified to pronounce on science as a whole – just as he used his position as a leading South African politician to project himself as a statesman of the world. Thus, on losing political office to Hertzog in 1924, Smuts devoted much energy to scientific affairs and to the S2A3 in particular. In his 1925 presidential address to the Association he sought to reorient science away from the 'habits of thought and the viewpoints characteristic of its birthplace in the northern hemisphere', pointing instead to the country's distinctive position in the southern African subcontinent and the southern hemisphere more generally. Employing a geological metaphor, Smuts posited Africa as the 'great "scientific divide" among the continents': a high exposed watershed (like the Witwatersrand) separating great drainage areas, where 'future prospectors of science may yet find the richest veins of knowledge'.[67]

Smuts began his 1925 address by describing the implications of the 'Wegener hypothesis', named after the German geophysicist whose *Origin of Continents and Oceans* (1915) pioneered the theory of continental drift. His use of Wegener's ideas – which were far from commonly accepted at the time – allowed Smuts to posit Africa as the 'mother-continent' of the southern hemisphere from which South America, Madagascar, India and Australasia originally split away or 'calved off'. In placing southern Africa at the centre of the southern hemisphere, it now became possible to correlate scientific developments across a range of disciplines in new and creative ways which might have far-reaching implications for 'universal science'.[68] In geology, for example, several Cape formations appeared to be mirrored by similar formations in India and South America; the pattern of mineral deposits, such as diamonds and coal, illustrated a similar symmetry.⁶⁹ Botany posed particularly interesting problems. For, whereas most South African flora was evidently of tropical origin, the southwestern Cape was characterised by a distinct temperate flora which could not – as was then widely believed – have derived from northern Europe. Instead, Smuts suggested, Cape flora might have come from ancient Gondwanaland, the continent which was now covered largely by the South Atlantic Ocean and which used to encompass much of Africa, Australia, India, South America and Madagascar. This theory could explain botanical affinities between South African flora and flora belonging to other countries in the southern hemisphere.⁷⁰ Continuing with the implications of the Wegener hypothesis for the understanding of climatic changes and meteorology, Smuts came to astronomy. Here he mentioned the outstanding pioneering work of Lacaille, Henderson and Gill, concluding that South African astronomy 'has the distinction of being responsible for the determination of both great astronomical standards of measurement – the distance of the sun and the distance of the fixed stars'.⁷¹

Finally, Smuts paid tribute to the developing field of human palaeontology, paying special attention to Raymond Dart's recent work on the Boskop and Taung skulls and their profound implications for the understanding of human evolution. Significantly, he lent support to Dart's novel and highly controversial claim that the Taung specimen (*Australopithecus*) represented a critical evolutionary 'missing link' in hominid development and he speculated that 'South Africa may yet figure as the cradle of mankind, or shall I rather, say, one of the cradles'. The idea that Africa might harbour some of mankind's earliest progenitors was a clear reference to Darwin but in speculating about 'cradles' in the plural form, Smuts might also have been gesturing towards polygenetic or multilinear theories of evolution. Whatever the case, Smuts certainly lent credence to the racialised form of anthropology that Dart did so much to promote. Thus, in arguing for South Africa's distinctive interest as a field of anthropological research, he ventured the opinion that 'Our Bushmen are nothing but living fossils' – analogous to the status of the country's cycads in the field of botany.

The notion that the aboriginal races of the country represented the end of an evolutionary line, that Bushmen could be likened to the indigenous flora and fauna of the country, and that they should be preserved primarily as evolutionary curiosities, had important implications for the development of racial science in South Africa and they were also reflective of racial attitudes during the segregationist era.[72] In this light, it is a matter of considerable irony – though by no means a contradiction – that Smuts prefaced his 1925 presidential address by affirming the ethnically inclusive spirit of the S2A3:

> In the Association both official languages of the Union enjoy equal privileges, and papers and addresses in either language are treated absolutely alike for purposes of publication or otherwise. It is the aim and object of this Association to bring together and unite all South Africans, irrespective of race and language, who are interested in the general scientific culture of South Africa.[73]

By 'irrespective of race and language' Smuts was obviously referring to English- and Afrikaans-speakers – not necessarily through a conscious act of excluding blacks, but simply because they would not have figured in this definition of 'South African'. Put more sharply, and considered in the light of the political prominence of segregationist legislation at this time, Smuts's appeal for unity amongst white South Africans reflected his unquestioning assumption of the need to maintain white supremacy and to deny African claims to common citizenship.

The 1929 visit

In 1929, at the invitation of the South African Association, the British Association returned to South Africa. This time there were more than 500 overseas visitors, funding having been secured from the South African government, individuals and commercial organisations based in Britain, as well as a contribution from the Rhodes Trust. The delegation was high-powered in scientific terms with some forty-four Fellows of the Royal Society attending, including its president, the atomic physicist Sir Ernest Rutherford.[74] Ten 'representative' scientists were invited to accept the hospitality of the Union. Thus, '[w]ith punctilious regard to the Union's racial traditions' three guests each were asked from England and Holland, the remainder coming from France, Austria, Italy and the United States.[75]

There were superficial resemblances with the 1905 meeting: as before, the official programme was split between Cape Town and Johannesburg (with optional excursions further afield), there were enthusiastic civic welcomes, extensive press coverage and evidence of considerable public interest in the proceedings. But the tone, context and intent of the 1929 meeting was markedly different from 1905. In the intervening quarter of a century, South Africa had been politically transformed. Instead of four separate British colonies under the overall control of a British High Commissioner, the ex-Boer war general J. B. M. Hertzog was now leader of a unitary state which had recently achieved the status of full sovereignty within the British Commonwealth. In the months before the Wall Street crash, the economy seemed buoyant. And the focus of white politics appeared to be turning away from tensions between English- and Afrikaans-speakers and towards the 'solution' of the 'native question'; indeed, only a few months before, Hertzog had fought – and decisively won – a general election on the issue of the 'black peril'.

By comparison with 1905, a new sense of national pride and self-confidence was widely evident. The composition of the nation was still a matter of dispute, but substantial consensus existed over two fundamental principles: that South Africa would continue to remain as an independent state within the Commonwealth, and that racial segregation between white and black was to be upheld and entrenched. Within these parameters the symbolic power of science and technology could easily be accommodated for it could be seen to exemplify the dynamism of a new society with independent standing in a wider community of nations. The British Association was keenly sensitive to the delicateness of the situation. If, in 1905, the British Association felt that it could afford to be imperious in keeping with its imperial role in the world, in 1929 it was now much more conscious of functioning within the emerging consensus politics of the Commonwealth and of its leading, but less elevated, position within the commonwealth of knowledge.

This spirit was evident from the outset. In his first address, the president of the British Association, Sir Thomas Holland, chose to quote from W. J. Viljoen's Afrikaans translation of Percy Fitzpatrick's *Jock of the Bushveld*, adding that 'the translation is in a language capable of expressing virile sentences'.[76] This gesture did not only play to the literary icon of South African Englishness; Holland's willingness to speak in Afrikaans struck an obvious chord with his audience, as did his tribute to Viljoen (who had died suddenly a few days before). A noted scholar of the Afrikaans language and leading educational administrator in the Cape as well as the Orange Free State, Viljoen was an exemplar of pragmatic moderation in the fraught arena of language and educational policy.[77] Thus, to reinforce the message of conciliation, Sir Ernest Rutherford made the point that the British Association gathering was a 'representative' one, noting that Sir Thomas was a Canadian while he himself was a New Zealander.[78] This formula was evidently successful for, on the Johannesburg leg of the journey, Holland again quoted in Afrikaans from *Jock*, adding that the book accurately 'reproduced the spirit of the veld'. And, at a mayoral reception attended by some 4,000 guests a couple of days later, the back of the city hall stage was hung with both the Union Jack and the new South African flag, below which the words 'Welcome/Welkom' were outlined in coloured lights.[79]

The 1929 meeting provided an ideal opportunity to reflect on South Africa's achievements since the beginning of the century. The fact that sectional meetings of the British Association could be held on the imposing new Groote Schuur campus of the University of Cape Town was a source of considerable civic pride; thus, the excellence of the university's facilities and architectural design (it was conceived by J. M. Solomon, a local Cape artist and follower of Herbert Baker, and built with the help of Edwin Lutyens) were widely remarked upon.[80] More broadly, science was explicitly called into the service of an inclusive white South African nationhood based on shared values of intellectual, cultural and material progress. This was frequently coupled with a firm commitment to international cooperation, especially in the context of the Commonwealth. Implicitly excluded were the extreme fundamentalism of Christian nationalism and, by a process of neglect and peripheralisation, the claims of African nationalism. The defining sentiments of this sense of moderate white nationhood were fittingly articulated by General Smuts's rising protégé, Jan Hofmeyr, the politician-scholar who would, in the 1930s and 1940s, come to personify the political hopes of the increasingly marginalised white liberal intelligentsia.

In welcoming the British Association to Cape Town on behalf of the South African Association (of which he was then president), Hofmeyr delivered a widely reported address entitled 'Africa and science' which the *Rand Daily Mail* referred to as brilliant, inspirational and visionary.[81] Hofmeyr began by recalling the remarkable intellectual progress which had taken place in South Africa since the previous visit of the British Association in 1905. He recorded the development of the university system, the growth of specialist and professional scientific societies, as well as the expansion in state-sponsored and applied scientific work.[82]

Hofmeyr's central theme was what he referred to as the 'South Africanisation' of science. Whereas in 1905 science in South Africa could be called 'exotic' in the sense that it depended on imported expatriates, its personnel was now, whether by adoption or birth, 'essentially South African'. But, although South Africans evinced great pride in their science because it was 'distinctively ours', Hofmeyr noted that there was 'nothing narrow about its South Africanism. Were it otherwise, it would have been false to the spirit of Science.'[83] South African science had succeeded in drawing the attention of the world to the country; it also had much to offer. Thus, Hofmeyr went on to express the view that the next task was even bolder, namely, to 'Africanise' science. This ambition was expressed in terms of a curious brand of intellectual sub-imperialism whereby whites, as the natural representatives of European civilisation in Africa, were bound to discharge their obligations to that great continent from their base in the south. In tones heavily reminscent of the evangelical nineteenth-century imperial mission Hofmeyr declared:

> It is by way of this Southern gateway that Science itself can most effectively be made to permeate Africa. And to you, having so come, to you, the ambassadors of Science, I present – Africa. It is Africa and Science, which, I would like to think, are to-day met together. Happy indeed should be the fruits of the mating.[84]

In closing, Hofmeyr described the 'development of Science in Africa, of Africa by Science' as a 'Promised Land'.[85] The optimism of this expansive African vision may have been misplaced, but it was not entirely without foundation as the concurrent opening in Pretoria of the Pan-African Agricultural Veterinary Congress as well as the International Geological Congress vividly illustrated.[86]

The sense of national pride and mission within which Hofmeyr conceived of South African science was enthusiastically taken up in the English-language press. It was reiterated by Smuts at the inaugural meeting of the Association in Johannesburg. On this occasion Smuts reminded his audience of the social and intellectual transformation which had taken place in South Africa since the previous visit of the British Association in 1905 to a country 'devastated and laid bare in a great war'. The reconstruction of the country, the achievement of political unity and the yet-to-be-achieved ideal of united nationhood, were matters of considerable pride. Smuts went on to say – with perhaps a permissible measure of exaggeration – that it was science that was largely responsible for recreating a 'new South Africa'. In areas such as mining, agriculture and manufacturing, science had provided the technology for overcoming animal and plant diseases and for exploiting natural resources in new and original ways. Whereas general members of the public had previously been sceptical about science, they were now appreciative and enthusiastic converts to its cause. Indeed, Smuts went so far as to suggest with double-edged irony – reflecting his suspicion of the mass technological age – that science had become 'the new magic ... the great intangible power, the imponderable which in the long run carries everything before it'.[87]

With these remarks Smuts touched various bases: his words lent added weight to Hofmeyr's Cape Town address, they endorsed the intrinsic association of white nationality with rational progress and material achievement, and (clever politician that he was) reminded his audience that he himself was both a broad-minded man of science and a central figure in the process of national conciliation and renewal.

Smuts also took the opportunity to underline the message of Sir Thomas Holland's presidential address in Johannesburg, appropriately on the mineral resources of the world and their importance as a factor in the maintenance of civilisation and world peace. Holland had stressed the role of applied science in the advance of material and cultural progress, linking the control of natural resources to international politics and arguing that if the British empire and the United States would jointly exercise a 'mineral sanction' on countries infringing the Kellog Pact, no war could last very long. His clear message, therefore, was that South Africa was a valued member of the Commonwealth and that the gold mining industry was of pre-eminent significance from both a domestic and an international perspective.[88] This idea was calculated to appeal to Johannesburg's industrial and business community, flattering their sense of importance and reminding them of their debt to science at one and the same time. It also served more generally to remind South Africans of their international obligations and responsibilities.

Holland's address can be seen as the culmination of a wider effort on his part to secure a broad base of support for scientific research. In a public debate on 'science and industry' held in Cape Town a week before, attention had been focused on the relationship between scientific research and economic prosperity. Stressing the beneficial effects of science to the community at large, Holland emphasised science as a democratic and popular, rather than an arcane and elitist, activity. Thus he sought to undercut both the condescending attitudes of pure scientists towards applied work, as well as the suspicion on the part of 'captains of industry' that scientific research was a useless if harmless activity undeserving of their backing.[89] In this regard Holland was not only addressing a South African constituency: like previous leaders of the British Association, he was alert to the need to reposition the organisation with regard to public and professional opinion back in Britain. In the 1920s, as Collins and Macleod have shown, the British Association found itself under renewed pressure. It had to respond to post-war public scepticism towards the idea that scientific research was an unqualified good; it needed to be seen to be serving the public interest; and, within the profession, it had to maintain the delicate balance between a commitment to pure and applied research – or, in other words, to justify the concerns and interests of the few, while satisfying the needs and understanding of the many.[90]

Just as South Africa's geopolitical significance within the Commonwealth was highlighted in speeches and comments at the 1929 meeting, so South African scientists were seen to be part of an international community and the production and dissemination of knowledge was understood as a reciprocal and cooperative process. Fittingly, therefore, the South African content of the 1929 conference was much more extensive and its profile significantly higher than in 1905. In fields ranging from zoology, botany and geography, to economics, anthropology, psychology and education, South African material was widely evident and discussed in some depth. As if to prove that South African scholarship could hold its own in international terms, mirroring its status as a nation amongst nations, General Smuts was invited to expound on his theory of 'Holism' at one of the prestigious set-piece events. His paper, grandly titled 'The nature of life', was given prominent coverage in the press and Smuts must have been gratified that scientists of the stature of J. S. Haldane, Wildon Carr, A. S. Eddington and Lancelot Hogben presented formal responses.[91]

In the wake of Dart's claims for the significance of *Australopithecus* as well as Robert Broom's lifetime of internationally recognised work on human and reptile fossils, anthropology and prehistory was accorded considerable attention. Several international luminaries in the field of prehistory were present in 1929, including Leo Frobenius, Henri Breuil, Gertrude Caton Thompson, John Myres and Henry Balfour. The latter, who was curator of the Pitt Rivers museum in Oxford and president of the anthropology section, paid tribute to the advances which had been made in South African prehistory since his first visit in 1899. And he stressed the national and international significance of South Africa's archaeological heritage.[92]

There were two especially keenly watched papers in the anthropological section, both with the potential to excite controversy. The talk by the German ethnologist and exponent of *kulturkreis* theory, Leo Frobenius, was well trailed in the press since he was expected to offer startling evidence for the existence of a great and civilised race which had apparently lived for over 7,000 years in the area between Great Zimbabwe and Lake Malawi, worshipping the stars and the moon. But Frobenius may have received advanced warning of Gertrude Caton Thompson's rather more prosaic findings about the origins of Great Zimbabwe; in the event his talk on the different styles and types of Bushman rock art – though suggesting Carthaginian influence in characteristic diffusionist mode – was presented in careful academic style and caused no sensation.[93]

The presentation of Caton Thompson's findings, by contrast, was no damp squib. As a respected archaeologist and Egyptologist, she had been commissioned by the British Association (with funding from the Rhodes Trustees) to settle the long-running popular and scholarly debate about the origins of the Zimbabwe ruins. Unlike Randall-MacIver's report to the 1905 meeting, Caton Thompson was firmly of the opinion that all aspects of the Zimbabwe civilisation were the results of indigenous African endeavour.[94] This finding, which unambiguously rejected theories ascribing the authorship of Zimbabwe to mysterious ancient visitors from the Near East, helped to puncture a key aspect of white settler mythology: that Africans were incapable of creating a significant material culture on their own. Needless to say, her views were not accepted by all. A determinedly dissenting note was sounded by the acclaimed anthropologist Raymond Dart whose fundamental commitment to cultural diffusionism caused him to launch an intemperate attack on Caton Thompson.[95]

Another controversial paper with a strong bearing on contemporary racial thought was E. G. Malherbe's discussion of 'Education and the poor white' which raised fears about the long-term capacity of the ruling white race to maintain its dominance. Drawing on the preliminary results of the landmark Carnegie Commission into poor whiteism (of which he was a leading member) the ambitious young Columbia-trained educationist shocked public opinion by referring to the poor white problem as a 'skeleton in our cupboard'. As a result of an unconscious feeling of vulnerability, Malherbe suggested, whites experienced a psychological 'inferiority complex' which led them to lay the blame for their own predicament on blacks. Mechanisms to protect the lowest ten per cent of whites from the upper fifty per cent of blacks would only serve to jeopardise the country's future.[96] Malherbe's address drew highly favourable comment by some and was widely seen as a courageous and authoritative intervention into an area of increasing social and economic concern. Dr C. W. Kimmins, president of the Educational Section, resorted to eulogy, and the *Cape Times* called it 'brilliant'.[97] Malherbe's mentor and colleague Fred Clarke, then professor of education at the University of Cape Town, wrote him a congratulatory letter in which he praised the young man for displaying 'Clear-headedness, courage & a genuine patriotism that was ready to face martyrdom'.[98]

By genuine patriotism, Clarke meant Malherbe's willingness to speak as a 'good South African' rather than an Afrikaner sectarian. Clarke's assessment of the attendant dangers proved correct when the Afrikaans press tore into Malherbe for betraying his Afrikaner loyalties. *Ons Vaderland*, mouthpiece of the Transvaal nationalists, accused Malherbe of 'professorial superficiality' and 'thoughtless stupidity'. The attack was ostensibly focused on Malherbe's methodology and the quality of his research, but the newspaper was clearly stung by the suggestion that whites were somehow psychologically fearful of blacks and highly sensitive about his calumnious airing in public of a problem which might reflect negatively on the standing of the Afrikaner *volk*.[99] *Die Burger* joined in with an editorial titled 'Pure nonsense', and *Die Volksblad* was likewise hostile.[100] Underlying the negative reaction lay an even deeper fear: that Afrikaner poor whites were becoming biologically and mentally enfeebled through eugenic deterioration.

Greatly concerned that the Carnegie enquiry was becoming dangerously politicised, a statement was communicated to the press by its leading investigators, J. F. W. Grosskopf, E. G. Malherbe and R. W. Wilcocks – all of whom were respected Afrikaner academics – stressing that their primary task was objective 'fact finding' and carefully affirming the commission's political neutrality. Malherbe was forced into a convoluted attempt to defuse the tension by distancing himself from the interpretation that had been placed on his words. This 'clarificatory' explanation was accepted by *Die Burger* which duly responded with an editorial acknowledging the importance of the commission's work and expressing confidence in its commitment to 'bring the facts to light'.[101] But *Ons Vaderland* remained uncompromisingly hostile and was especially concerned at the negative impression that would be created by Malherbe's presentation of so-called 'facts' to the British Association as well as to his American backers.[102]

Political controversy thus intruded into the 1929 meeting, but it did not overwhelm the event. In the many reflections and tributes to the success of the 1929 meeting, both the British and the South African Associations expressed themselves well pleased with the way in which things had gone. A special meeting to reflect on the proceedings was convened by the Royal Empire Society in December. Here, the editor of *Nature*, Sir Richard Gregory, reiterated the modern utilitarian role of science in promoting imperial cooperation, economic development and human well-being. In an age 'when nearly all the problems of development of natural resources and guidance of life are scientific, yet the control of the factors of progress is in the hands of administrators who have no first-hand knowledge of science', it followed that science should rank in importance with the 'administrative and fighting services' and be regarded 'as an indispensable part of the machinery of government and not as a luxury to be dispensed with in times of financial stringency.'[103] In South Africa, a genuine mood of buoyant confidence in the country's achievements was evident. The *Cape Times* captured the spirit when it proclaimed in an editorial on 'Savants and servants':

> When last the British Association met in Cape Town its delegates were the guests of the Government of the Cape Colony, then a separate State in the South African group of colonies. But science knew no political boundaries, and in 1905, as in 1929, it held its meetings in the chief centres of what now constitutes the Union, and in doing so it contributed its share to that strong sense of South African unity which was then beginning to take shape and form.[104]

The idea that science knows no politics was of course a convenient fiction. Aside from Malherbe's intervention into the causes of poor whiteism, one of the striking features of the 1929 programme was the number of papers dealing directly or indirectly with racial science and eugenics. Amongst these, one might mention the contributions of visitors like R. Ruggles Gates on 'racial crossing' and H. J. Fleure on 'racial drifts'. Even more directly pertinent to South African conditions were the raft of papers on physical anthropology and human origins, the heated public debate on the origins of Great Zimbabwe, as well as papers dealing with comparative racial intelligence quotients and the purpose of native education.

I have argued elsewhere[105] that the dispassionate qualities of science were frequently invoked at this time in regard to the desire to find a 'solution' to the 'native question'. In Hofmeyr's 1929 address, for example, he maintained that science had an important role in 'determining the lines along which white and coloured races can best live together in harmony and to their common advantage'.[106] The appeal to – and appeal of – scientific objectivity can also be seen as a counterpart to the (disingenuous) wish of leading politicians to prevent the 'native question' becoming an issue in party politics. If blacks were in the process of being excluded from common citizenship via the landmark segregation bills then under parliamentary consideration, justification for their exclusion was in part founded on the idea that African culture was incompatible with the values of Western rationality and natural progress. Science could therefore be used both to evaluate Africans' rights as citizens, and also to constitute white citizenship and nationality.

As the Malherbe incident illustrated, the terms upon which white nationhood would be constituted was becoming increasingly bitterly contested. A resurgent Afrikaner nationalism operating both within and beyond the realm of parliamentary politics sought to define South African nationality in ways that went much further than the 'two streams' policy of Prime Minister Hertzog. The objective of radical nationalists who sought ultimately to establish a Boer republic without any trace of the imperial connection was not yet fully articulated. But the tendency was registered in the only significantly discordant note sounded in reaction to the British Association's 1929 visit – that of the Afrikaans press.

The reception granted to the British Association by *Die Burger*, the mouthpiece of Cape Afrikaner nationalism, in 1929, was cool rather than overtly hostile. Notably, the newspaper chose to emphasise the European rather than British profile of the visitors, paying particular attention to the presence of individuals like Professor Freudenberg of the University of Heidelberg. In subdued tones of selfless patriotism, the paper expressed willingness to learn from their overseas guests, noting that the visit was in South Africa's best interests and that it would improve the international status of the country.[107] It was on this basis, too, that *Die Kerkbode*, the influential organ of the Dutch Reformed Church, came to terms with the Association's visit. *Die Kerkbode* was uncomfortable, however, with the fact that there was no explicit theological dimension to the meeting, but comforted itself with the observation that religion was an intrinsic part of everyday intellectual life and existence. Exception was taken, however, to Hofmeyr's opening address in which he referred to the exciting possibility that Africa might, as Darwin predicted, turn out to be 'the scene of Nature's greatest creative effort', namely, the evolution of mankind. Quoting Hofmeyr's account of recent palaeontological and anthropological discoveries, *Die Kerkbode* therefore expressed disappointment at his acceptance, in defiance of scriptural orthodoxy, of evolutionist theory. Notwithstanding this complaint, the Dutch Reformed Church organ was pleased to welcome the scientists: it warmed to the promise that science could help to advance South Africa's agriculture and industry, and noted with satisfaction Thomas Holland's efforts to speak in Afrikaans.[108]

If Afrikaner opinion-formers were inclined to accept the British Association, they would only do so on their own terms. They were thus extremely sensitive to any appropriation of the visit by supporters of Smuts's South African Party, or to anything which could be interpreted as a slight on the Afrikaner volk. It is therefore unsurprising that the British Association provided a convenient subject for party political sniping. In anticipation of the British Association's arrival the *Cape Times* (a staunch supporter of Smuts's South African Party) published a crude cartoon by Whyndam Robinson. This depicted two of Hertzog's ministers, F. W. Beyers and D. F. Malan, dressed as cavedwellers, with Beyers chewing on a bone (or meat) inscribed 'hatred'.

The caption read: 'Those members of the British Association especially interested in relics of the Stone Age will not have far to look.'[109] This provocation did not go unnoticed. *Die Burger* responded in an editorial by the nationalist ideologue A. L. Geyer entitled 'scandalous'. Addressing itself to the learned visitors, the paper generously sought to assuage any embarrassment they might suffer; the *Cape Times*, it assured them, was a jingoistic newspaper which could not best claim to represent British values and culture. Nor should the *Cape Times* be seen as a yardstick of South African culture. Afrikaners – whether English- or Afrikaans-speaking – were ashamed that the visitors should encounter a tasteless newspaper which venomously attacked the nationalist government that had been responsible for inviting them. For those guests accustomed to reading accounts of 'Dutch' racial hatred in British newspapers, the *Cape Times'* cartoon revealed where the source of such sentiment was truly to be found.[110] In sum, Geyer's editorial skilfully sought to invert accusations of race hatred, to distance the British Association from Smuts's South African Party and to align Afrikaner nationalism with South Africanism. This tactic was endorsed in an article by C. J. Langenhoven, the noted Afrikaner poet and critic, who claimed the high-ground of cultural and intellectual achievement for the National Party, and attacked those who viewed Smuts as a great world statesman who suffered domestically from an unappreciative nation of 'backvelders' and blinkered race-haters.[111]

Notably, the hostility of the Afrikaner press was directed to specific issues and did not extend to a direct attack on the British Association itself or, indeed, on western scientific traditions. Pride in South African scientific achievement was widely shared and actively courted. Anger was instead directed inwards and targeted towards moderate South African Party-supporting Afrikaners like Malherbe, Smuts and Hofmeyr, whose ease with the international community of scholars was in itself perceived as a threat.

Conclusion

The visits of the British Association to South Africa in 1905 and 1929 highlight marked shifts in the relationship between South Africa and the imperial metropole. In 1905 the British Association was greeted by a newly constituted country whose experience of war had left it economically shattered and politically fragmented. Discussions about future political cooperation between the four British colonies were only beginning but, aside from the centralised authority of the colonial administration, there existed few national institutions out of which South Africa could be fashioned as a meaningful entity. At least as far as the anglophone establishment was concerned, the joint scientific meeting of the British and South African Associations offered an excellent opportunity to proclaim the virtues of peaceful cooperation and reconstruction. Such an event could demonstrate the connection between Britain and its southern African colonies, as well as the internal links between those colonies. And it would do so in a manner that was neither official nor politically provocative: under the guise of the universality of science and the search for objective truth, the universality of the British empire and the transcendent virtues of British values could be implicitly affirmed. By expressing Englishness (or, more properly, Britishness) in the neutral language of universal progress, prosperity and security, the risk of giving political offence was minimised.

In the Victorian era the notion of civilisation had served as a key discursive justification of British imperialism, building variously on evangelical Christianity, racial supremacy and social evolutionism. But the civilising ideal carried unwanted historical baggage in the shape of John Philip, Exeter Hall and the humanitarian tradition. These aspects of the civilising project were acceptable neither to Afrikaners, nor to those forward-looking English-speaking South Africans who sought an answer to the country's problems in political reconciliation between Boer and Briton and racial segregation between black and white. One promising way out of this ideological impasse was to secularise and neutralise the idea of civilisation, stripping it of its moralistic and culturally specific overtones and recasting it (and imperialism more generally) in the guise of scientific progress and technological prowess. Portraying government policies in this light was precisely what Milner and his followers sought to do in the post-war reconstruction period. Equally, because science was an apparently neutral and objective activity committed to the realisation of the common good, it was also readily adopted by politicians like Botha and Smuts to aid the creation of an independent and unitary state under Afrikaner leadership.

Science and technology can therefore be said to have played a notable role in the process of political reconciliation between Boer and Brit. Indeed, in the years after Union the creation of a range of scientific and educational institutions came to symbolise the growing confidence of a new South African nation which, though independent, was nevertheless still firmly tied into the wider imperial network. However, the rise of Afrikaner nationalism in the immediate post-Union era under the leadership of General Hertzog, increasingly challenged the ideology of South Africanism championed by the governing

South African Party. With the electoral victory of Hertzog's National party in 1924 this challenge became more overt. Hertzog's 'two-stream' ethnic policy, coupled with his quest to assert South Africa's parity within the empire, was initially accommodated within the newly developing white commonwealth. Nevertheless, when the British Association revisited South Africa in 1929 the changed political milieu could be measured in terms of discernible shifts in the relationship between imperial and colonial science. South African patriotic pride and confidence were registered by a demonstrably enhanced scientific and technical ability to master the African environment (and, in the case of anthropology, Africans themselves). For politicians like Smuts and Hofmeyr science was therefore both a source of national pride in South Africa's achievements as well as visible proof of its claims to international status: it was, in this sense, both a tangible expression of, and a metaphor for, commonwealth.

In the sphere of domestic politics, political moderates hoped that scientific rationality would triumph over ethnocentric prejudice and emotion (and for many of the same figures a belief in scientific rationality extended to efforts to 'solve' the 'native question'). As the proceedings of the 1929 meeting revealed, however, this optimistic vision of the dispassionate benefits of scientific research was tainted by the nationalist sensitivities which surfaced in discussions about poor-whiteism and evolutionist theory, and in stereotypes of Boer backwardness. But although such disputes questioned the extent to which political reconciliation was an accomplished fact, they did not overly detract from the broader success of the British Association in developing a supra-white sense of national identity. Rather, they pointed to intensifying debates about the position of the Afrikaner volk within the body politic of South Africa as well as its broader relationship to the outside world. These debates were, however, not only about the terms of inclusion within white South African identity; increasingly, they referred to the basis upon which blacks would be excluded from common citizenship. And here, too, science played a conspicuous role.

Notes

1 My thanks to Jocelyn Alexander, William Beinart and Libby Robin for their comments on earlier versions of this chapter.
2 *The Times*, 2 September 1905.
3 See photograph album, A17 fol., British Association for the Advancement of Science, Historical Papers, University of the Witwatersrand.
4 Archive of the British Association for the Advancement of Science (BAAS) Bodleian Library, Oxford. Box 205, *Engineering*, 28 July 1905, p. 119.

A COMMONWEALTH OF SCIENCE

5 O. J. R. Howarth, *The British Association for the Advancement of Science: A retrospect 1831–1931* (London, 1931), pp. 128–30; *The Times*, 16 August 1905; BAAS box 203, Donald Currie to Professor Dewar, 11 June 1903.
6 *The Times*, 16 August 1905; 2 February 1905.
7 BAAS box 420; *Morning Post*, 30 October 1905.
8 *Star*, 29 August 1905.
9 See e.g. BAAS box 198, *Glimpses in Natal*.
10 *Star*, 28 August 1905; *The Times*, 17 August 1905.
11 James Starke Browne, *Through South Africa with the British Association* (London, 1906), p. 275. It was precisely because of a perception that the 1884 meeting in Montreal would amount to 'a glorified picnic of important men of science, who could have no serious purpose in visiting Canada', that objections to overseas meetings were registered in England. See Howarth, *The British Association*, p. 121. See also *Star*, 28 August 1905 (editorial).
12 This paragraph is drawn from R. Macleod, 'Introduction', in R. Macleod and P. Collins (eds), *The Parliament of Science: The British Association for the Advancement of Science 1831–1981* (Northwood, 1981), pp. 17–19, 32–4. See also A. Gale, 'Science at the margins: the British Association and the foundations of Canadian anthropology, 1884–1910', Ph.D. thesis, University of Pennsylvania, 1986, pp. 67, 187–9.
13 M. Worboys, 'The British Association and empire: Science and social imperialism, 1880–1940', in Macleod and Collins (eds), *Parliament of Science*. The classic analysis of social imperialism remains B. Semmel's *Imperialism and Social Reform* (London, 1960).
14 Gale, 'Science at the margins', pp. 22–3.
15 Address by Sir David Gill to the S2A3, in *Report of the South African Association for the Advancement of Science: First Meeting Cape Town 1903* (Cape Town, 1903), p. 17.
16 *Ibid.*, pp. 18–19. BAAS box 201, Gill to secretary British Association, 5 August 1902; Gill to Garson, 9 March 1903; Gill to Silva White, 25 January 1904.
17 Address by Sir David Gill to the S2A3, p. 36. cf. the words of Gill's fellow astronomer John Herschel who did pioneering work at the Cape between 1834 and 1838 and who was also a leading light in the early years of the British Association (serving as president in 1845): 'Let selfish interests divide the worldy; let jealousies foment the envious; we breathe a purer Empyrean air. The common pursuit of truth is of itself a brotherhood.' Cited in R. Macleod, 'Retrospect: The British Association and its historians', in Macleod and Collins (eds), *Parliament of Science*, p. 4.
18 *Morning Post*, 30 October 1905, in BAAS box 420.
19 *The Times*, 16 August 1905.
20 *The Times*, 31 August 1905.
21 *Star*, 2 September 1905.
22 *Ibid.*
23 *Star*, 28 August 1905 (editorial); *The Times*, 23 August 1905; *The Times*, 14 September 1905.
24 *The Times*, 23 August 1905 (speech by Professor Darwin); *The Times*, 6 September 1905; *Star*, 2 September 1905 (speech by Professor Darwin). *The Friend*, 4 September 1905.
25 BAAS box 201, Gill to Silva White, 14 March 1905; Gill to Silva White, 28 March 1905; James Stewart to Gill, 24 March 1905.
26 *The Times*, 28 August 1905; Browne, *Through South Africa*, p. 69.
27 *Star*, 5 September 1905; *The Times*, 6 September 1905; Browne, *Through South Africa*, pp. 127–8; *Bloemfontein Post*, 6 September 1905.
28 *Diamond Fields Advertiser*, 5 September 1905.
29 *The Times*, 13 September 1905.
30 H. T. Montague Bell (ed.), *Addresses and Papers Read at the Joint Meeting of the British and South African Associations for the Advancement of Science Held in*

South Africa 1905, 4 vols (Johannesburg, 1906). See e.g. BAAS box 202, William Cullen to David Gill, 24 May 1905. Cullen took exception to Gill's suggestion that the editors were having to 'cadge' for papers, but admitted that abstracts were not coming in quickly and that some sub-standard papers 'have managed to slip in'.

31 A. C. Brown (ed.), *A History of Scientific Endeavour in South Africa* (Cape Town, 1977), pp. 417, 442–3; B. Warner, *Astronomers at the Royal Observatory Cape of Good Hope* (Cape Town, 1979), pp. 104–5. These surveys were massive enterprises. Gill's project to measure, by triangulation, the 30th east meridian (the longest measurable arc of meridian) stretching from South Africa to Norway, was only completed in 1955. See also Howarth, *The British Association*, p. 130; D. Gill, 'On the origin and progress of geodetic survey in South Africa, and of the African arc of meridian', in *Report of the Seventy-Fifth Meeting of the British Association for the Advancement of Science: South Africa 1905* (Cape Town, 1906), pp. 228–48.

32 Presidential address by T. Reunert, *Report of the South African Association for the Advancement of Science: 3rd and 4th meeting 1905–6* (Cape Town, 1906), p. iv; Review of F. H. Hatch and G. S. Corstorphine, *The Geology of South Africa* (London, 1905), in *Star*, 29 August 1905.

33 A. C. Haddon, 'Anthropology', in *Report of the Seventy-Fifth Meeting of the British Association*, pp. 511–27.

34 *Ibid.*, p. 304. Notably, MacIver's paper was the only anthropological contribution with reference to South Africa printed in the main section ('Reports on the state of science') of this volume.

35 *The Times*, 2 September 1905; Pim's address was reported at length in *Star*, 1 September 1905. For the significance of Pim's writings, see S. Dubow, *Racial Segregation and the Origins of Apartheid in South Africa, 1919–36* (London, 1989).

36 Richard C. Jebb, 'Educational Science', in *Report of the Seventy-Fifth Meeting of the British Association*, pp. 597–605; for comments on the significance of this address, see M. Boucher, *The University of the Cape of Good Hope and the University of South Africa 1873–1946* (Pretoria, 1974), p. 67.

37 *The Times*, 2 September 1905: Professor Darwin, in his closing address, referred to the 'number of important papers bearing on South African matters'.

38 See e.g. *Star*, 28 August 1905 (editorial): 'We flatter ourselves, moreover, that this visit has been looked forward to no less eagerly by them than by the Colony'; also *Star*, 2 September 1905 (editorial): 'It is also something gained if the visit has enabled them to gather – though few of them were likely ever to have entertained doubts on the point – that the British element in South Africa is neither worse nor better than the members of the stock from which it has sprung'; *Bloemfontein Post*, 2 September 1905 (editorial): 'We are indeed very sensible of the honour and privilege of entertaining, though it be for a moment, the representatives of old-world learning though we regret that our own advancement makes us worthy of being little else than a pied a terre between greater centres.'

39 S. M. Naudé and A. C. Brown, 'The growth of scientific institutions', in Brown (ed.), *A History of Scientific Endeavour*, p. 70.

40 N. J. Dippenaar (ed.), *Staatsmuseum 100* (Pretoria, 1992). When the British ousted the Kruger government from Pretoria in 1900 it was rapidly renamed the Pretoria (and later, again, the Transvaal) Museum.

41 *Report of the South African Association for the Advancement of Science: First Meeting... 1903*, p. 9.

42 The S2A3 was cautious not to be seen as in any way interfering with the premier scientific status of the South African Philosophical Society which dated back to 1877 and of which Gill was also president in 1903. The relationship between the two organisations is comparable to that existing between the Royal Society of London and the British Association. The superior position of the Philosophical Society, though never really in doubt, was confirmed when it received the Royal Charter in 1908 and changed its title to the Royal Society of South Africa.

43 'Report of the Council for the period ended May 2nd 1903', in *Report of the South African Association for the Advancement of Science... 1903*, p. 517; B. J. F. Schonland, 'The South African Association for the Advancement of Science, its past and future', *South African Journal of Science* (hereafter *SAJS*), 49:3–4 (1952), 61. Membership was drawn from the whole of South and southern Africa.
44 Presidential address by T. Reunert, in *Report of the South African Association for the Advancement of Science... 1905–6*, p. i.
45 Address by Sir David Gill to the South African Association for the Advancement of Science, in *Report of the South African Association for the Advancement of Science... 1903*, p. 36.
46 *African Monthly*, 2:8 (1907), 227.
47 Macleod and Collins (eds), *Parliament of Science*, pp. v, 17.
48 D. Denoon, *A Grand Illusion* (London, 1973), pp. 68–9. See also J. Krikler, *Revolution from Above, Rebellion from Below: The agrarian Transvaal at the turn of the century* (Oxford, 1993), pp. 66, 76 and ff.
49 See W. Beinart, 'Agricultural science, conservation and nationalism: H. S. D. du Toit in South Africa', paper presented to conference on 'Science and Society in Southern Africa', University of Sussex, 1998.
50 B. C. Jansen, 'The growth of veterinary research in South Africa', in Brown (ed.), *A History of Scientific Endeavour*, pp. 166–70; F. Stark (ed.), *Pretoria: 100 Years* (Pretoria, 1955), pp. 161–5. In 1980, Theiler was the first recipient of the S2A3's South African Medal.
51 D. M. Joubert, 'Agricultural research in South Africa: An historical overview', in Brown (ed.), *A History of Scientific Endeavour*, pp. 266–8.
52 The national survey arose out of a merger between the Geological Commission of the Cape Colony and the Geological Survey of the Transvaal. The Geological Survey of Natal and Zululand ceased to exist some years before Union. See Dippenaar, *Staatsmuseum 100*, p. 62.
53 R. F. H. Summers, *A History of the South African Museum, 1825–1975* (Cape Town, 1975), p. 92; N. Fowler, 'A history of the Albany Museum, 1855–1958', unpublished MS, 1968, Albany Museum, pp. 183–4.
54 E. Percy Phillips, 'A brief historical sketch of the development of botanical science in South Africa and the contribution of South Africa to botany', *SAJS*, 27 (1930).
55 H. Bolus, 'The native flora of South Africa and its preservation', *The State*, 2:7 (1909), 105–6. See also, 'The proposed national botanic garden', *Report of the Eighth Annual Report of the South African Association for the Advancement of Science: Cape Town 1910* (Cape Town, 1911), pp. 421–3.
56 H. H. W. Pearson, 'A state botanical garden', *The State*, 5:5 (1911).
57 H. H. W. Pearson, 'A national botanic garden', *Report of the Eighth Annual Report ... of the South African Association*, p. 54.
58 R. H. Compton, *Kirstenbosch: Garden for a nation* (Cape Town, 1965), pp. 33, 50.
59 *Ibid.*, p. 31.
60 Although Kirstenbosch was founded as a National Botanic Garden, this status was only uneasily maintained. Cape dignitaries provided a great deal of impetus for its creation and it was more closely linked for research purposes to the University of Cape Town than to a government department (like that of agriculture). At a meeting on 8 March 1912 attended by many Cape notables it was resolved to establish 'a National Botanic Garden within the Cape Peninsula'. In 1923 a new herbarium was established under the aegis of the department of agriculture at Pretoria as an embryo 'Kew'. This initiative was regarded as a snub to Kirstenbosch. Local Cape opinion was indignant at what it saw as regional competition and an unwelcome instance of state centralisation. See Compton, *Kirstenbosch*, pp. 43, 80–1; also D. P. and E. McCracken, *The Way to Kirstenbosch* (Cape Town, 1988), chs 11 and 12.
61 These figures should be treated as indicative rather than entirely accurate; there are inconsistencies in the sources from which they have been drawn: J. H. Hofmeyr, 'Africa and science', *SAJS*, 26 (1929), 4; E. G. Malherbe, Memo on

'Educational development in the Union 1905–1930', Killie Campbell Library, Durban, KCM 56973 (305) file 427/7, p. 5; E. G. Malherbe, *Education in South Africa vol. II: 1923–75* (Cape Town, 1977), p. 727.
62   Boucher, *The University of the Cape of Good Hope*, p. 124; R. P. B. Davis, 'University reconstruction in the Cape Colony', *African Monthly*, 3:18 (1908); J. Edgar, 'Union and the university question', *The State*, 3:4 (1910); 'The case for a national university', *The State*, 4:1 (1910).
63   M. Boucher, *Spes in Ardus: A history of the University of South Africa* (Pretoria, 1973); E. G. Malherbe, *Education in South Africa vol. 1: 1652–1922* (Cape Town, 1925); H. Phillips, *The University of Cape Town 1918–1948: The formative years* (Cape Town, 1993).
64   UG [Union Government publication] 33-'28, *Report of the University Commission* (Pretoria, 1928), paras 48, 105–6 and 'summary of conclusions'.
65   E. A. Walker, *The South African College and the University of Cape Town 1829–1929* (Cape Town, 1929), p. 78.
66   See S. Dubow and S. Marks, 'Patriotism of place and race: Keith Hancock on South Africa', paper delivered Keith Hancock symposium, Australian National University, Canberra, 1998.
67   J. C. Smuts, 'South Africa in science', *SAJS*, 22 (1925), 3–4.
68   Ibid., pp. 4–5.
69   Ibid., pp. 5–6.
70   Ibid., pp. 6–7.
71   Ibid., pp. 14–15.
72   See e.g. R. J. Gordon, *The Bushman Myth* (Boulder, 1992); S. Dubow, 'Human origins, race typology and the other Raymond Dart', *African Studies*, 55:1 (1996).
73   Smuts, 'South Africa in science', p. 1.
74   Howarth, *The British Association*, p. 143; *The Times*, 22 July 1929.
75   *Cape Times*, 11 March 1929.
76   *The Times*, 23 July 1929.
77   *Cape Times*, 20 July 1929 (editorial).
78   *The Times*, 23 July 1929.
79   *Rand Daily Mail* (hereafter *RDM*), 23 July 1929; *RDM*, 25 July 1929.
80   See e.g. *Cape Times*, 27 July 1929 (editorial); Phillips, *The University of Cape Town*, p. 8.
81   *RDM*, 23 July 1929 (editorial). See also *Cape Times*, 23 July 1929 (editorial); *The Times*, 5 August 1929 (editorial).
82   Hofmeyr, 'Africa and science', pp. 2–5. Cf. press reports of Hofmeyr's talk in, e.g., *The Times*, 27 March 1929.
83   Hofmeyr, 'Africa and science', pp. 6–7.
84   Ibid., p. 9.
85   Ibid., p. 18.
86   *RDM*, 3 August 1929 (editorial); *Cape Times*, 29 July 1929. Some of the meetings of the Geological and Agricultural sections of the British Association were held in Pretoria in conjunction with these other international conferences.
87   *RDM*, 1 August 1929. See also *RDM*, 2 August 1929 (editorial).
88   *RDM*, 1 August 1929.
89   *RDM*, 24 July 1929; 25 July 1929 (editorial); 2 August 1929 (editorial); 5 August 1929 (interview with Holland). Among the participants in this discussion were Richard Gregory, D'Arcy Thompson and Daniel Hall. They covered subjects such as science's contribution to industry, fisheries and soil fertility. See *Cape Times*, 24 July 1929.
90   Macleod, 'Retrospect' and P. Collins, 'The British Association as public apologist for science, 1919–1946', in Macleod and Collins (eds), *Parliament of Science*, pp. 5–6, 211–15.
91   *The Times*, 26 July 1929; W. K. Hancock, *Smuts, vol. 2: The fields of force* (Cambridge, 1968), pp. 190–1. Hogben, in typically pugnacious fashion, was less respectful than most. He dismissed Smuts's metaphysical approach, maintaining

that Holism could only be treated seriously if it could be presented in a form capable of being tested.
92   *The Times*, 2 August 1929. Balfour also visited South Africa in 1905 as part of the British Association.
93   *RDM*, 20 July 1929; 31 July 1929; *Cape Times*, 1 August 1929.
94   *Star*, 2 August 1929; *RDM*, 3 August 1929.
95   For more on Dart and the racial paradigm he promoted, see S. Dubow, *Scientific Racism in Modern South Africa* (Cambridge, 1995); also Dubow, 'Human origins, race typology and the other Raymond Dart'.
96   *Cape Times*, 24 July 1929. See also E. G. Malherbe, 'Education and the poor white', *SAJS*, 26 (1929).
97   *Cape Times*, 24 July 1929; 24 July 1929 (editorial); 25 July 1929.
98   E. G. Malherbe papers, KCM 56973 (47), Fred Clarke to E. G. Malherbe, 24 July 1929. Clarke later became professor of education at the University of London.
99   *Ons Vaderland*, 26 July 1929 (editorial).
100 *Die Burger*, 26 July 1929 (editorial); *Die Volksblad*, 30 July 1929 (editorial).
101 *Die Burger*, 30 July 1929.
102 *Ons Vaderland*, 7 August 1929 (editorial).
103 BAAS box 261, 'Science and the empire' address given by Sir Richard Gregory on 3 December 1929.
104 *Cape Times*, 22 July 1929 (editorial).
105 Dubow, *Racial Segregation*; Dubow, *Scientific Racism*.
106 Hofmeyr, 'Africa and science', p. 16.
107 *Die Burger*, 19 July 1929; 20 July 1929 ('Suid-Afrika en die wetenskap').
108 *Die Kerkbode*, 31 July 1929 ('Die "British Association"').
109 *Cape Times*, 17 July 1929.
110 *Die Burger*, 18 July 1929 ('Skandelik', editorial by A. L. Geyer).
111 C. J. Langenhoven, 'Aan stille waters', *Die Burger* (I am unable to find the precise date).

CHAPTER FOUR

'For the public benefit': livestock statistics and expertise in the late nineteenth-century Cape Colony, 1850–1900[1]

Dawn Nell

> The people of this land have asked through their representatives in Parliament for information regarding its resources, and the Government relies upon its Officers satisfying this public want for the public benefit.[2]

Livestock and crop enumeration could evoke strong sentiments in the Cape Colony in the late nineteenth century. While certain sectors of the colony's inhabitants viewed statistics as an indispensable aspect of 'modern' life and put pressure on the civil service to provide more reliable statistics, other sectors of the population viewed enumeration with suspicion. This chapter looks at the tensions surrounding agricultural statistics in the context of 'Progressive' and 'anti-Progressive' white politics. During the second half of the nineteenth century a new approach to agriculture – characterised by an emphasis on scientific investigation and high levels of capital investment – was sweeping the colony and, with the backing of legislators, threatened to put an end to certain of the agricultural practices relied upon by many of the colony's farmers.

Knowledge, observation and enumeration

Throughout the nineteenth century, enumeration had become a hallmark of the Progressive ethos, as first scientists, then statesmen, began to insist on rationality as the basis for understanding the world. Far-reaching shifts in epistemological procedures had already begun to emerge in the eighteenth century; empiricism, with its emphasis on the precise observation and description of phenomena according to explicit methods, was one of the methodological manifestations of this. Numerical information facilitated the observation of data, and statistics rapidly became one of the defining features of scientific discourse.[3] It was at this point that nature writing – known throughout the eighteenth century as 'statistical description' – was expelled from science, and statistics came to be defined more narrowly as purely *numerical* description.[4]

Michael Adas has argued that it was the general Victorian penchant for 'statistical reductiveness', reinforced by advances in mass printing and in graphic and statistical representation which encouraged the replacement of less rigid forms of description by empirical observation.[5] The significance of this new emphasis was twofold. Firstly, it provided the basis for a sense of intellectual superiority. Adas quotes John Seeley – champion of British nineteenth-century imperialism – as arguing that British knowledge was simply 'better trusted and sounder'. Secondly, it instilled confidence in the ability of observers to acquire complete knowledge of any region through description and enumeration.

By the early nineteenth century scientific discourse was being actively employed in the description and analysis of society with a firm conviction that statistics and scientific rationality could provide the answers to societal problems. This view was encapsulated by J. R. McCulloch, a vigorous promoter of statistics in Britain during the 1830s, who argued that 'the accumulation of minute and detailed information from all parts of the country would, at length, enable politicians and legislators to come to a correct conclusion as to many highly interesting questions that have hitherto been involved in the greatest doubt and uncertainty'.[6] John Brewer has observed that the collection of statistical information had begun to extend into new areas of enquiry throughout the nineteenth century. Whereas state bureaucracies had initially collected statistics on trade, they now began to include resources and potential resources conceptualised and collated on a national scale. European states, he says, had started to envisage national wealth in new ways.[7] As Anne Godlewska points out with reference to Napoleonic France, '[o]ne of the characteristics of the state's growing awareness of its realm had . . . been an evolution in the understanding of what constituted national wealth and thus what merited the attention of the government'.[8]

Partly as a reflection of this expansion in state interests and activities, statistics came to be closely associated with the centralised and 'rational' bureaucracies that were developing during this period. These bureaucracies became preoccupied with the search for 'scientific' solutions to the nation's problems. By the nineteenth century governance in modern states was characterised by an insistence on objective and rational decision-making. This had a significant effect on the nature of the civil service, as civil servants were now required to develop administrative expertise of which numerical proficiency was an important part. Bernard Silberman has argued that the rationalisation of state bureaucracies developed as a response to political uncertainty which had led to the development of bureaucratic systems in which there was an emphasis on the utilisation of rules of decision-making that were presented as being objective.[9] Statistics were particularly well suited to the requirements of 'rational' state bureaucracies. Theodore Porter has argued that because of the highly structured nature of the discipline, statistics facilitates communication that can be accurately translated beyond boundaries of locality and community and is, therefore, an integral facilitator of centralised administration.[10] Statistical description had its greatest significance, however, in its claims to objectivity. In a context where administrators were expected to make decisions based on objective and rational deduction, statistics could provide the basis for state action.

While the format of the state's informational requirements was to a certain extent determined by its administrative needs, it was also important that the information should not be seen purely as part of the state's administrative agenda. The collection of information in specific formats was part of the process by which information-gathering by the state was constructed as knowledge.[11] By the nineteenth century, the prerogative of legislators and administrators to make decisions about a country or colony was premised largely upon the bureaucracy's role as collector of information, which allowed the state to be regarded as the custodian of 'true knowledge'.

Statistics, politics and progress

While they frequently disagreed on what the statistics meant, self-appointed 'Progressive' commentators in the nineteenth-century Cape Colony – including administrators, farmers and parliamentarians – constantly urged the importance of comprehensive and accurate enumeration.[12] Progress, they argued, would not be possible until there was a body of 'accurate' statistical information upon which to base decisions. In the minds of such men the colony's development was something that was in serious need of attention. In the face of growing competition from other parts of the empire, the Cape appeared to be lagging behind in agricultural development. The principal enthusiasts for enumeration in the colony were, therefore, the self-conscious 'Progressives', described by Tony Kirk as 'a rising generation of self-made men, whether British or Afrikaners ... among its ranks, the "progressive" sheep and cattle ranchers, the village store keepers and artisans, the accountants, attorneys, newspaper editors and professional men', and their representatives in parliament.[13] Agriculture, they argued, needed urgently to be modernised, and compiling statistics on livestock and crop production was seen as the first crucial step to achieving this. In 1899, in the midst of concerns about the quality of wool being produced in the colony, the Chief Inspector of Sheep urged farmers to take note of the pressures of global competition:

> If the Cape Colony was the only wool-producing country in the world, if the wool manufacturers were dependent upon our export of the raw material for their annual supplies, there would exist a lamentable absence of healthy competition, and farmers in the colony might possibly offer this as an apology for the inferiority of the article they produce. But this is not the case. The rivals we once had have outstripped us in the march of progress, and there are other competitors, who are surely and rapidly leaving us in the rear ranks. At the present time our wool, when offered on the London markets, is looked at askance, buyers shrug their shoulders, and like the Pharisee pass by on the other side ... The natural result is that the prices offered decline, and will still further decline, until we, by our own supineness and lack of energy, allow our wools to be shouldered altogether from the markets of the world.[14]

Merchant capital was one of the major driving forces behind progressive agriculture in the Cape in the nineteenth century, with merchants investing in agricultural improvements, from breeding stock to farm implements and processing equipment, and encouraging farmers to improve their produce.[15] It was not without reason that the Commissioner of Public Works in 1882 said that 'The best Minister of Agriculture is the merchant, who competes with others for the farmer's produce, who introduces new machinery, and leads him to improve the staple so as to get command of the world's markets.'[16] With the growth of the colony's press and the granting of first, Representative Government in 1853, then Responsible Government in 1872, the voice of Progressives had come to dominate colonial society. Commercial men and wealthy farmers came to control parliament and many of them prided themselves on their forward-looking outlook. Central to their self-image as decision-makers was their access to information and, as such, statistical information went hand in hand with political authority. Statistics on livestock and crop production were at the centre of debates on livestock disease legislation, the development of the railway network and even the nature of political representation in the Cape.[17] Statistics formed the basis of 'true knowledge', which, so far as they were concerned, was what provided legislators with the authority to make decisions regarding the colony and its inhabitants. John X.

Merriman, the leading parliamentarian and member for Namaqualand, said himself in 1894 that Members of the House 'were there to teach their constituents from the better knowledge they were able to obtain'.[18]

While a number of members would attempt to uphold the views of their rural constituents who objected to certain state agricultural policies and interventions, the political dichotomy constructed between 'Progressives' and 'non-Progressives' made their task almost impossible. Both principal political groupings in the Cape parliament in the late nineteenth century identified themselves as 'Progressive'. The Afrikaner Bond, political voice of Afrikaner farmers from the 1880s onwards, was strongly progressive and, until the shock of the Jameson Raid in 1895, formed an alliance with the English-speaking mercantile class under the leadership of Cecil John Rhodes. This accommodation engendered a bitter rift within the Bond as Afrikaner sheep farmers were disappointed at their party's refusal to oppose anti-Scab legislation. The Scab Act was designed to put an end to the 'wasteful' farming practices which still prevailed in much of the Cape interior and which were blamed for the spread of the disease. The intention was that 'inefficient' farming practices should be replaced by an altogether more scientific, 'rational' approach. Consequently, the disagreement over the Scab Act was ultimately articulated as a conflict between 'Progressives' and 'anti-Progressives'.

Statistics helped to swing the balance in favour of the former, ensuring that the perspectives of Progressive farmers and state experts be regarded as expert *knowledge*, while relegating the understandings of other farmers to the status of mere *superstition*. In a discussion on the Cape's railway system in 1890, Thomas Theron, MP for Richmond, complained of the ease with which the opinions of many of the colony's inhabitants could be dismissed. This was possible, he observed, because 'it was said that the country people judged in ignorance, and that their public meetings were worthless'.[19] The response of another member of the House seemed only to confirm Theron's concerns:

> [T]he political education of the electors of that district are sorely in want of more information before they were able to pass an opinion upon the railway question ... Their infancy in politics is demonstrated by there being no other man competent to preside at a public meeting and explain the position of public affairs than the parson.[20]

The prominence of rationality and expertise in the Progressivist ethos made the construction of a public discourse of scientific knowledge on the colony an urgent project. Rational decision-making was itself regarded as a sign of progress and rational deduction was not considered possible without 'accurate' information. Before regular census enumeration was introduced in 1865, Progressive commentators were concerned that the Cape compared unfavourably with other British colonies in the collection of statistics.[21] In the early 1890s, Progressives were pressing for even more regular statistics on livestock and crop production, complaining that census information was neither comprehensive nor accurate enough to form the basis for decision-making. In his address to Parliament at the start of the 1890 sitting, the Governor had noted that '[a] serious drawback to much-needed legislation of a progressive character has been the absence of statistical information'.[22] Similarly, in 1892, a Member of the House of Assembly argued that '[t]he country was greatly in need of such returns. There could be no true legislation except upon the basis of statistics'.[23]

Progressive commentators in the Cape made a clear connection between the colony's potential for progress and its ability to attract settlers and capital, and statistics constituted an important framework for comparison as the colony competed for attention with other parts of the empire. So central was such data to the colony's image that when there was discussion on holding a census in 1901 a number of members of the House of Assembly argued against it on the grounds that it would portray an unduly negative picture of the colony so soon in the wake of the South African War. Josias Hoffman, MP for Paarl, argued that taking a census so soon after the disruption of the war, 'would be bad policy ... for the small returns would certainly damage the Colony's credit in Europe'.[24]

Such reticence notwithstanding, the publication of statistical information was considered crucial in the construction of a public understanding of the colony and its prospects. Livestock and crop statistics were not only routinely published in promotional material for overseas audiences or in the annual government Blue Books, but also appeared regularly in popular publications such as the *Cape of Good Hope Almanac* and the *Cape Monthly Magazine*. As Jean du Plessis has argued, the aim of publishing such information was to assist in constructing the idea of the colony as a 'political, economic and social whole, rather than a conglomerate of disparate identities'.[25] Leading colonial politicians, commentators and scientists regarded the publication of this so-called 'useful information' as a central component in the education of the populace to an understanding of progress. Published information on livestock numbers, livestock disease and production levels, they believed, would point to standards against which the performance of individuals could be measured and, as such, it was hoped, they would be driven to pursue improvements for themselves.

Scientific agriculture and statistics

As part of the search for ways of improving agricultural production along scientific lines, the need to compare levels of productivity provided a central motivation for the collection of agricultural statistics. The rise of the discipline of economics and the growing focus of the natural sciences on solving 'practical' problems, including problems of national progress, gave further impetus to statistical methods. Underlying the development of scientific agriculture, says Charles Rosenberg, was the contemporary understanding that 'productivity' was the essence of an index to 'progress'.[26] Progressive farming was oriented towards export markets and sought to combine scientific theory, practical experience and sound commercial acumen. In 1892 the Department of Agriculture's *Journal* outlined the lineaments of 'progressive' farming when it said that 'Agricultural Science [was] the necessary complement of agricultural practice' and added that '[n]ow-a-days the farmer must keep his books and make his calculations as well as the merchant or manufacturer'.[27] At another point the *Agricultural Journal* quoted Walter Long, President of the Board of Agriculture in Britain: 'They [farmers] had got to learn how they could put the best article on the market at the lowest price they could to yield a profit. Was it not to science that they must turn if they were to solve these problems satisfactorily?'[28] The proponents of 'progressive' farming were as convinced of the 'ignorance' and 'backwardness' of many of the colony's farmers as they were of the 'benefits' of scientific agriculture. For the Progressives, science provided the only way of understanding the world and as such was of 'obvious' application in improving farming methods. A contemporary observer, commenting on the need for agricultural schools, urged that practical farming experience was not enough; men needed 'to learn all the scientific laws'.[29] Scientific farming was predicated upon a firm belief in the universality of knowledge and the collection of statistical information was part of the project to integrate the Cape into a broader framework of scientific knowledge on agriculture. Statistics operated in this context to facilitate the dialogue between agricultural practitioners and scientific experts and was regarded by Progressives as an essential element of the state's role as the supporter of scientific agriculture and, by extension, colonial progress. Science helped to turn agriculture into a modern vocation; it was what separated farming as a business from farming as 'mere subsistence'. When a farmer and member of the Legislative Assembly reflected upon the changes in agriculture in the last decade of the nineteenth century, he commented on the growing professionalisation of agriculture, noting that in the past our boys have gone in for the liberal professions because they thought it more fashionable. Farming was considered as below the work of a doctor or lawyer or surveyor and others. During the last four or five years, that feeling has changed, and people now look to agriculture as being fashionable, or even more fashionable than some of the liberal professions.[30]

Progressive or professional farmers distinguished their activities from those of other farmers by the extent to which scientific method had been infused into their farming practice. An important characteristic of progressive agriculture was, therefore, the detailed measurement and recording of information on agricultural production. Reflecting on the debate in the *Agricultural Journal* on the exact parameters of an acre, a reader argued that to farm using 'happy-go-lucky' estimates and not to consider the costs, and specifically the labour costs, of production, was 'business without book-keeping, and one can only guess the relation between the cost of raising a crop and the profit it brings in'.[31] Farmers who did not operate using accurate and reliable information were operating at best in the realm of guesswork and at worst in the realm of superstition. A letter from the Stellenbosch Agricultural School in 1893 urged the necessity of standardised measurement in agriculture. The author complained that many a farmer had 'little knowledge of weights, measures of capacity and square-measure, but merely [spoke] of a 'morgen' as x strides long by x strides wide and measures his corn by the bag, his potato crop by the basket or *muid* and other articles by the load.'[32] The apparent lack of understanding of, or interest in, accurate measurement and recording by many farmers, was regarded as an obvious barrier to the improvement of farming techniques. One of the problems in discussions on irrigation, for example, was the confusion over the meaning of 'an inch of water'. The colony's chief hydraulic engineer referred to it as 'the meaningless expression used by farmers when talking of water flowing through an aperture' and urged the necessity of standardising such terms.[33] State scientific experts also repeatedly insisted on greater accuracy in measuring livestock dipping tanks and dips. The Scab Commission in 1894 blamed the ineffectiveness of dipping by farmers in the Cape interior on their haphazardness in taking measurements and went so far as to propose that farmers who did not calibrate their dipping tanks accurately should face a fine of £20.[34] Standardised measurement was so central to notions of scientific farming because it was this that facilitated the integration of local information and knowledge into a universal framework of scientific understanding.

Progressive farmers had a keen interest in the work of scientific experts because of the advice they could offer but were also very quick to criticise them and their institutions if they failed to prove concrete results – that is, cost-effective ways of improving production.[35] Yet, even while questioning the work of state experts on occasion, such farmers were nevertheless convinced of the merits of science in improving agricultural productivity and they conducted their own experiments and came up with their own theories. The *Agricultural Journal* regularly carried letters and articles with suggestions and observations by its readers.[36] The enthusiastic response of many farmers to the Department of Agriculture's request for statistics on crop yield is a further reflection of this desire to find scientific explanations to agricultural problems.[37]

As farmers developed their understandings of agricultural practice based largely on their own observations, agriculture was an environment in which the opinions of scientific experts were particularly vulnerable to criticism. Rosenberg argues with regard to the work of agricultural scientists in America in the nineteenth century, that farmers had to be trained to accept the opinion of the scientist.[38] Nonetheless, a common educational background and a shared understanding of what progress entailed provided the foundation for a dialogue between state experts and progressive farmers. State experts, in acknowledging the specific nature of farming conditions in the Cape, welcomed observations by 'intelligent farmers'.[39] While disagreement was not uncommon, the dialogue was always undertaken in the discourse of scientific rationality and with a confidence that science could provide 'true' solutions.

Opposing enumeration

Like observers elsewhere, many inhabitants of the Cape had doubts about the ability of statistics to provide a meaningful reflection of conditions in the colony. A satirical article published in the *Uitenhage Times* in 1865 alluded to a lack of confidence in the capabilities of enumerators and, more broadly, questioned the relevance of enumeration itself. The writer assumed the voice of a semi-literate enumerator, asking questions like: 'What's yer age?' 'Air you marrid? and if so how do you like it? How many children have you, and do they sufficiently resemble you as to proclude the possibility of their belongin to any of yer nabers?' 'Did you ever have the masels, and if so, how many?' 'Do you know any Opry singers, and if so, how much do they owe you?' 'How many chickens hav you, on foot and in the shell?'[40] The author of the article not only poked fun at the alleged lack of education of many of the enumerators, but also caricatured the state's statistical queries in terms of an obsession with meaningless data.[41]

'FOR THE PUBLIC BENEFIT'

The proponents of statistics did what they could to counter claims of irrelevance and pointed to what they claimed was the increasing accuracy of enumeration techniques. The colonial press was firmly behind the Progressive cause and was engaged in an ongoing campaign to educate the colony's inhabitants about the significance of enumeration. An article in the *Cape Argus* in 1875, for example, complained that census enumeration was,

> greeted in many quarters as a joke, and false or inaccurate particulars are likely to be given... Everyone possessing any common sense must be well aware of the value of statistics, and none are more important than the number of inhabitants in a country, the quantity of land under cultivation, how many flocks and herds are in existence, and the like.[42]

For farmers not to see the usefulness of statistics was, however, a more serious issue as a lack of enthusiasm for statistics had, by the last few decades of the century, come to be regarded as a reflection of a broader opposition to progress itself; something that, in the minds of Progressives at least, the colony could ill afford. While Barnabas Shaw in 1841 may have celebrated a Liliefontein farmer, 'who possessed three or four hundred sheep and goats [and who] could never count further than twenty', Progressive commentators would have found such an approach to farming more difficult to entertain later in the century.[43]

By the latter half of the nineteenth century, livestock enumeration was no longer directly linked to taxation. Yet the warnings repeatedly issued by the Department of Agriculture suggest that many of the colony's farmers remained suspicious of livestock counts. Even by 1904, and despite imposing a £5 fine for furnishing inaccurate information, Department of Agriculture officials were still complaining that farmers were underenumerating their livestock.[44] There were a variety of reasons for this lack of interest in statistics, ranging from concerns about individual privacy, to cynicism about how the information would be used.[45] (In many cases, it seems there was a lingering concern that livestock enumeration might still be linked to taxation.)[46] For many farmers the construction of a body of scientific knowledge for the colony and the associated expansion in the authority of state scientific experts went hand in hand with unwelcome state intervention in agriculture.

The Cape pastoral interior had been characterised by transhumant pastoralism throughout the eighteenth and nineteenth centuries and, while it had been largely replaced by intensive livestock farming by the late nineteenth century, large areas of the colony, particularly in the north, were still home to significant numbers of farmers engaged in semi-nomadic pastoralism. The continued existence of this form of farming was the source of a great deal of concern for, implicit in understandings of 'colonial progress', was a belief that nomadic pastoralism was essentially anti-Progressive and had to be replaced by sedentary agriculture. When the Surveyor General outlined his vision for development in the northern Cape in 1860, he said:

> I have no doubt that the alienation of these Trekvelden to private owners is a mere question of time, which will be forced on the good sense of the divisional councils and the public by the contrast which their pasture will present, in their unimproved condition, with their unopened, badly-used fountains and water-holes, and wasteful occupation, when colonial progress should have pushed enterprising land-owners, the makers of dams, the builders of houses, and the cultivators of the soil, in greater numbers to the neighbourhood of the mere nomadic grazier.[47]

Simply put, semi-nomadic 'trekfarmers' were regarded by the proponents of progressive farming as ignorant and their farming practices were dismissed as inefficient and backward.[48] By the 1880s, the progressive farming lobby had begun to make a concerted attack on trekfarming. Trekfarmers were blamed for the lack of 'colonial progress' in these areas because of their reluctance to make 'improvements' to the land, while their insistence on moving their stock was blamed for the spread of livestock disease. The proponents of progressive agriculture characterised the unwillingness of these farmers to adopt scientific farming methods as a sign of overwhelming ignorance. The Colonial Veterinary Surgeon summed up the frustration of state experts in 1885 when he said: 'There are a great many minds amongst the farming community in this Colony formed after the model of a certain apostle's, whom no amount of logic will convince... No amount of reasoning will make converts of such men, and I may be only wasting paper in continuing the discussion.'[49] When the Scab Commission carried out its investigation between 1892 and 1894, the Commissioners were struck by the disjuncture between their own understanding of the disease and the views held by many of the farmers they encountered. The report of the Commission explained that it had been necessary to 'minutely cross-examine witnesses... for the purpose of elucidating statements, often made without sufficient forethought, or to hear out preconceived notions which had no foundation in fact, and which on minute examination were consistently found to be untenable'.[50] The lack of interest shown by many farmers in livestock enumeration was regarded as symptomatic of this same 'irrationality' and 'ignorance'.

Like the proponents of progressive agriculture, these farmers believed that knowledge was gained through experience and observation. The difference, however, was that they were not as convinced of the universality of science as were many Progressives. They were suspicious when state scientists appeared with their theoretical approach to farming. As Mordechai Tamarkin has noted, one of the features of trekfarming communities in the late nineteenth century was their reluctance to accept that knowledge about farming could be gained from books.[51] In 1882, the Commissioner of Public Works pointed out that '[t]he farmers are not very prone to take the advice of theoretical strangers. They ridicule the idea of experts from Europe being able to show them anything.'[52]

While progressive farmers deemed farmers in the remote interior as 'backward', the latter defended their knowledge of the region and how best to farm it. They argued that their farming practices were adapted to the peculiarities of their situation and designed to take best advantage of local environmental vicissitudes. Farmers of the interior were also acutely aware of how their methods differed from those of other parts of the colony or the empire and were not prepared simply to accept that what scientific experts said worked in one region would necessarily work in their own. They argued that they had a very clear idea of the conditions in which they were operating and, at a time when many of the measures advocated by state experts were meeting with indifferent success, many of these farmers were reluctant to believe that 'progressive' or 'expert' measures were necessarily advantageous. Thus, when one north-western Cape farmer was asked why he had moved his stock contrary to Scab Act regulations, his response was simply that 'he [knew] the conditions of his veld better than the Inspector'.[53] In 1896, when the Member of Parliament for Calvinia complained that the Scab Act was not practicable in the northern Cape, he defended the local knowledge of the region's farmers; it was their right not to be convinced by the authority of state experts such as the Colonial Veterinary Surgeon:

> Those farmers had taught themselves to farm; they knew their sheep and their veld, and no-one could teach them what they ought to do with their sheep. Their grievances were genuine, and nothing could persuade them to believe the contrary. The advocates, doctors and other learned men of the House got their information from books; the farmers learned by experience, and how could such people know what was good for farmers and what farmers were capable of doing. They – the advocates and doctors – failed to see that what could be carried out in Australia could not necessarily be carried out in South Africa.[54]

Conclusion

The latter half of the nineteenth century had been witness to a bitter contest over agricultural production by white farmers within the Cape Colony. Self-proclaimed 'progressive' farmers with the support of most of the colony's legislators regarded scientific farming as the colony's only hope. The scientific agriculture which they advocated was infused with an emphasis on precise – preferably numerical – information and the application of scientific or rational methods. A close correlation between 'scientific' and 'progressive' was crucial to the self-perception of the colony's ruling classes as they devoted their energies to overcoming what they perceived as almost insurmountable obstacles to agricultural progress in the colony.

By the beginning of the twentieth century, statistics were so central to modern understandings of expertise and governance that they had conclusively become part of the outlook of all the ascendant political parties in the colony. The so-called 'Anti-Progressives' were increasingly marginalised, both by ongoing state intervention in agricultural production and by the growing strength of a bureaucratic system premised upon 'rationality'. State-sponsored agricultural instruction and greater access to formal education reduced the 'anti-Progressives' to an insignificant minority and ultimately helped secure the state's role as collector of information and champion of agricultural development amongst white farmers. This process was linked to the development of white nationalism and the politics of Union; in later decades the state would increasingly turn its attention to rectifying 'unscientific' agricultural practices in the rural African reserves.

Notes

1 This chapter is based on research undertaken for my Master's thesis at the University of Cape Town in 1998 entitled 'You cannot make the people scientific by Act of Parliament: the state, farmers and livestock enumeration in the Northwestern Cape, c. 1850–1900'. I hereby acknowledge the financial support of the South African Centre for Science Development.
2 Cape Archives Repository (hereafter CAR), AGR, vol. 2, fol. 7/1, Circular no. 9 of 1894.
3 David Spurr, *The Rhetoric of Empire: Colonial discourse in journalism, travel writing and imperial administration* (Durham and London, 1993), pp. 21–6. Foucault's *The Order of Things* (New York, 1973) offers a detailed account of the new classificatory systems that came into being during this period. Statistics were, however, not always an accepted element of scientific discourse. Mary Poovey argues that statisticians in the early nineteenth century had to work hard to prove that statistics were free from *a priori* assumption and could be regarded as a science; see 'Figures of arithmetic, figures of speech: the discourse of statistics in the 1830s', *Critical Inquiry*, 19 (Winter 1993).
4 Theodore Porter, *Trust in Numbers: The pursuit of objectivity in science and public life* (Princeton, NJ, 1995), pp. 5–18.

'FOR THE PUBLIC BENEFIT'

5 Michael Adas, *Machines as the Measures of Men: Science, technology, and ideologies of Western dominance* (New York, 1989). The phrase 'statistical reductiveness' is, however, Richard Altick's, *Victorian People and Ideas: A companion for the modern reader of Victorian literature* (New York, 1973).
6 Cited in Poovey, 'Figures of arithmetic', p. 263.
7 John Brewer, *The Sinews of Power: War, money and the English state, 1688–1783* (London, 1989), p. 224.
8 Anne Godlewska, 'Napoleon's geographers (1797–1815): Imperialists and soldiers of modernity', in Anne Godlewska and Neil Smith (eds), *Geography and Empire* (Oxford, 1994), p. 37. The significance of this flow of information for colonial administrations has been documented in Thomas Richards, *The Imperial Archive: Knowledge and the fantasy of empire* (London and New York, 1993) and C. A. Bayly, *Empire and Information: Intelligence gathering and social communication in India, 1780–1870* (Cambridge, 1996).
9 Bernard Silberman, *Cages of Reason: The rise of the rational state in France, Japan, the US and Great Britain* (Chicago and London, 1993).
10 Porter, *Trust in Numbers*, pp. 1–7.
11 David Ludden argues that the scientific, objective basis of the state's knowledge constituted the increasingly dominant component of colonial discourse by the late nineteenth century; see 'Orientalist empiricism: Transformations of colonial knowledge', in Carol A. Breckenridge and Peter van der Veer (eds), *Orientalism and the Postcolonial Predicament: Perspectives on South Asia* (Philadelphia, 1994).
12 For example, the request of the Stutterheim Farmers' Association for regular statistics on Scab disease; *Agricultural Journal of the Cape of Good Hope* (hereafter *Agric. Jnl*) (1 March 1900), 291.
13 Cited in Jean du Plessis, 'Colonial progress and countryside conservatism, an essay on the legacy of Van Der Lingen of Paarl, 1831–1875', MA thesis, University of Stellenbosch, 1988, p. 57.
14 *Agric. Jnl.* (19 January 1899), 86.
15 Saul Dubow, 'Land, labour and merchant capital in the pre-industrial rural economy of the Cape: The experience of the Graaff-Reinet District (1852–1872)', *Communications*, no. 6 (1982); Hermann Giliomee, 'Western Cape farmers and the beginnings of Afrikaner nationalism, 1870–1915', *Journal of Southern African Studies*, 14:1 (1988), 38–63; Wayne Dooling, 'The decline of the Cape gentry, 1838–c. 1900', *Journal of African History*, 40:2 (1999), 215–42.
16 C.2-'82. *Report of the Select Committee appointed by the Legislative Council to Consider and Report upon the appointment of a Minister of Agriculture*, p. 1.
17 The publication of the 1865 census results added new fuel to the ongoing dispute between the eastern and western provinces of the colony over political representation; see *Cape Argus*, 27 June 1865; 29 June 1865; 25 July 1865. In 1890, livestock and crop statistics were employed by critics of the government's scheme for the extension of railways to bring the proposal into question; see *Cape Hansard*, 1890, pp. 4–5, 34–5, 55–69, 73–7.
18 *Cape Hansard*, 1894, p. 165.
19 *Cape Hansard*, 1890, p. 71.
20 Ibid.
21 G.21-'60. *Correspondence Relative to a Proposal From the Government of Australia for Taking a Census of the British Colonies Generally, in 1861*.
22 *Cape Hansard*, 1890, p. 1; see also J. X. Merriman's comments, p. 55.
23 *Cape Hansard*, 1892, p. 38.
24 *Cape Hansard*, 1899, pp. 391–2.
25 Du Plessis, 'Colonial progress', p. 74.
26 Charles Rosenberg, *No Other Gods: On science and American social thought* (Baltimore, 1997), p. 141.
27 *Agric. Jnl*, 4 (28 January 1892), 170–1.
28 *Agric. Jnl*, 17 (20 December 1900), 799–800.
29 A.1-'1909. *Report of the Select Committee on the Agricultural College*, p. 28.

30 A.1-'1909. *Report of the Select Committee on the Agricultural College*, p. 12.
31 *Agric. Jnl*, 12 (12 May 1898), 593.
32 CAR, AGR, vol. 2, fol. 7/1, 6 May 1893. A *muid* is a Dutch measure of capacity, about 90 kg.
33 G.27-'86. *Report of the Hydraulic Engineer for the Year 1885*, p. 10.
34 G.1-'94. *Report of the Scab Commission, 1892–1894*, p. 23.
35 State spending on scientific initiatives and institutions regularly came under criticism in discussion on the Department of Agriculture's annual vote. See for example the debates over the work of the Bacteriological Institute in Grahamstown; *Cape Hansard*, 1895, p. 508 and 1898, p. 173. See also the discussion on state trout-breeding experiments; *Cape Hansard*, 1899, p. 583; and for a similar discussion on the closure of the state-run stud farm; *Cape Hansard*, 1890; p. 153.
36 A reader in 1895, for example, argued that the decline in sheep numbers in the old sheep-farming districts was due to overstocking and added that 'I know that the Veterinary surgeons attribute the disease prevailing in these parts to other causes.' *Agric. Jnl*, 8 (30 May 1895), 266; see also *Agric. Jnl*, 16 (15 February 1900), 247.
37 *Agric. Jnl*, (3 November 1892), 224. There is also a large amount of correspondence on these returns in CAR, AGR, vol. 2, fol. 7/1.
38 Rosenberg, *No Other Gods*, p. 148.
39 A comment attached to a letter to the Department of Agriculture submitting statistics on wheat yield in 1893 said, 'This is the first reply to our circular from an intelligent man, I suppose the results may be fairly accurate'; CAR, AGR, vol. 2, fol. 7/1; Merriman to Secretary for Agriculture, 19 January 1893.
40 *Uitenhage Times and Farmers' Courant*, 30 June 1865.
41 Concerns about the literacy levels of enumerators were sufficient in 1865 for provision to be made for administrative assistants to assist enumerators who were illiterate or only partially literate; A.70-'65. *Instructions issued to the Respective Civil Commissioners relative to the employment of Enumerators and Clerical Assistants in taking the Census of the Colony*.
42 *Cape Argus*, 4 March 1875.
43 Barnabas Shaw, *Memorials of South Africa* (London, 1841), p. 77.
44 CAR, CSS, vol. 4/1/1, ref. 401, 18 February 1893, and CSS, vol. 4/1/2, July 1894.
45 For complaints about the intrusion of census enumerators see CAR, CSS, vol. 4/1/1, ref. 401, 18 February 1893; CSS, vol. 4/1/2, 18 December 1893 and *Cape Argus*, 1 April 1865, p. 3. Fears about the possibility of an uprising in response to a census just after the South African War are discussed in *Cape Hansard*, 1900, pp. 288–90, 391–2.
46 The final report of the 1965 census noted that the returns for livestock and crop production were 'somewhat unsatisfactory'. The report stressed the difficulty of conducting a census in a 'thinly-peopled colony' and pointed out that the colony's inhabitants viewed demands for statistical information as 'necessarily preced[ing] measures for increased taxation' (G.20-'66. *Report of the Census*, p. 5).
47 G.30-'60. *Copy of Correspondence with the Divisional Councils of Certain Divisions in which 'Trekvelden' exist on the Subject of the Future Disposal of these Lands*, Surveyor General to Colonial Secretary.
48 The Civil Commissioners' reports contained in the annual *Blue Books of Statistical Returns* consistently pointed to the lack of 'progress' in the Cape pastoral interior. Extracts from the reports of 1859 provide some reflection of what seems to have been commonly held views of state officials on these districts. The Civil Commissioner of Clanwilliam complained that 'while other districts of the colony have been rapidly progressing in agriculture and commerce, it appears that Clanwilliam has not advanced one step'. The Civil Commissioner of Calvinia said that pastoralism in his district 'gives little or no trouble, promotes indolence and consequent want of energy... The farmers are thus rendered incapable of undertaking any improvements.' *Blue Book*, 1859, p. JJ3.
49 G.26-'86. *Report of the Colonial Veterinary Surgeon for 1885*.
50 G.1-'95. *Report of the Scab Commission, 1892–1894*, p. 2.

51 Mordechai Tamarkin, 'Flock and volk: Cape Afrikaner sheep farmers in the mid 1890s: Between ethnic calling and the call of economic survival', paper presented at the 16th Biennial Conference of the South African Historical Society, 6–9 July 1997. William Beinart has similarly noted that one of the main difficulties facing state experts in combating livestock disease, was in persuading farmers that the diseases were known to science, 'Vets, viruses and environmentalism at the Cape', in Tom Griffiths and Libby Robin (eds), *Ecology and Empire: Environmental history of settler societies* (Edinburgh and Pietermaritzburg, 1997).
52 C.2-'82. *Report of the Select Committee Appointed by the Legislative Council to Consider and Report upon the Appointment of a Minister of Agriculture*, p. 1. See also *Agric. Jnl* (28 January 1892), 170.
53 G.11-1909. *Report on the Scab Acts and their Administration in the North-western Districts of the Cape Colony*, p. 6.
54 *Cape Hansard*, 1896, p. 227.

CHAPTER FIVE

A mania for measurement: statistics and statecraft in the transition to apartheid[1]
Deborah Posel

This chapter is an exploration of the power of numbers in apartheid South Africa. Apartheid statecraft – the ways in which the powers, responsibilities and methods of the state were envisaged – included aspirations of totalising modes of racialised knowledge. Bureaucrats engaged in rituals of often absurdly detailed quantitative measurement in their continuous efforts to count and classify the population. The apartheid state (particularly the Department of Native Affairs, the vanguard of apartheid policy-making) imagined the possibility of its own omniscience in the form of all-encompassing, racially disaggregated and elaborately cross-referenced statistical data, which quantified the extent of a host of 'problems' confronting the state and the manner of their 'solutions'. But visions of power seldom coincide with the more complex realities of its exercise. This chapter's exploration of the apartheid state's epistemology and practice of social measurement reveals both the blustering grandiosity of its imaginings of power, and some of the more mundane, at times farcical, realities of its governance.

My immediate focus is on the connections between the apartheid state's capacity to count and to control. More broadly, I am concerned with the impact of what were deemed to be 'modern' modes of political rationality on the emergence and development of apartheid. To avoid essentialising the idea of modernity, it should be seen as a set of processes which are always rooted in specific times and places and which do not unfold uniformly along a single historical trajectory. Yet the concept of modernity is also intended to articulate historical and geographical interconnectedness. I argue that the mania for measurement illustrates ways in which the apartheid project was understood and represented as a 'modernising' one, invoking many of the assumptions, expectations and norms typical of modern states elsewhere in the world. At the same time, it was the enthusiasm for designing more 'modern' modes of governance that produced apartheid's aberrant uniqueness.

Modernity and measurement: some conceptual and historical connections

Political compilations of quantified information about people, populations and property go all the way back to ancient times, though popular and governmental enthusiasm for quantifiable knowledge underwent a sea-change during the nineteenth century. According to Ian Hacking, 'between 1820 and 1840, there was an exponential increase in the number of numbers that was being published',[2] reflecting new epistemological assumptions about the power of measurement. Increasingly, measurement became the hallmark of intellectual authority and the very condition of knowledge itself. 'At the end of the century, no-one could dissent from the saying of the physicist Lord Kelvin, "that when you can measure what you are speaking about, you know something about it; when you cannot measure it... your knowledge is of a meagre and unsatisfactory kind".'[3]

As Foucault and others have argued, the new-found intellectual power of statistics was closely interwoven with the emerging 'technology of power in a modern state'.[4] There are interesting variations in the genesis of this connection, but 'by the late nineteenth century, statistics was absorbed into a mode of thought and argument in both the social sciences and public policy in all Western polities, irrespective of their political diversity'.[5] The reasons for this confluence lie in the close affinities between statistics as a mode of representing social realities and modern representations of the role and character of the state.

According to Foucault, the idea of the modern state was a product of the discourse of 'governmentality' – a set of characteristically modern assumptions which dominated political thinking in Europe from the late eighteenth century. The roots of these discursive shifts, says Foucault, go back to the sixteenth century, during which the very idea of government was problematised in a new way.[6] Subverting the notion of a divine cosmological order which defined the place and purpose of the political sovereign, government was now understood as 'an autonomous rationality',[7] with a normative logic of its own. By the nineteenth century, he claims, the 'early modern' idea of the 'art of government' had largely given way to the idea of a 'science of government'.[8] As Colin Gordon points out, 'this modern governmental rationality is simultaneously about individualising and totalising: that is, about finding answers to the question of what it is for an individual, and for a society or population of individuals, to be governed or governable. Different ways of posing and answering these questions compete and coexist with one another.'[9] For the purposes of this chapter, it is the 'totalising' dimensions of modern governance which are of primary interest – the ways in which modern governmentality represented the tasks of government as the technically expert management of 'problems of population'.[10] The purpose of government evolved as the effort to promote the wealth, welfare and longevity of the population, a shift which was in turn made possible by the proliferation of statistical knowledge. In showing that 'population has its own regularities ... that the domain of population has a range of intrinsic, aggregate effects'[11] which can all be quantified, statistics became an integral part of the process of constructing 'governable realms of reality'.[12]

These ideas of the 'science of government' in turn shaped the institutional characteristics and ambitions of modern states. Zygmunt Baumann argues that modern states are distinguished, 'first and foremost', by their 'centralisation of social powers previously localised'.[13] A tendency also nascent in Europe from the sixteenth century (in the emergence of large territorial monarchies), the process of centralisation was itself bound up with the logic of 'population': if government was a process of managing large units of population, understood impersonally and *en masse*, then the locus of power was situated most appropriately at the centre, the vantage point from which the problems of population could be conceptualised in unified, aggregate terms.

Centralising the exercise of power thus also changed its character. Enacted on a large canvass of the population as a whole, and with opportunities for a panoptic view of the full extent of it, government could also become a more purposefully rational, carefully planned and expertly managed set of techniques. Large, administrative bureaucracies therefore became the engines driving the power of the modern state. As Max Weber has argued, norms of legal-rational authority have undergirded modern bureaucracies, in the conviction that bureaucratic efficiency and competence rest in the application of instrumentally rational procedures applied uniformly and impersonally across all cases.[14]

The centralisation of power also magnified the complexity of government, in ways which shaped the production of bureaucratic expertise. With one central state entrusted with the responsibility and power for managing all units of the population, bureaucracies allowed for the disaggregation and compartmentalisation of manifold and complex tasks into discrete, manageable stages, each one the province of an official with an expertly detailed, narrowly specific knowledge of that particular area. In terms of the logic of modern statecraft, the role of the bureaucrat was to contribute the reasoned objectivity of a dispassionate, skilled and expertly knowledgeable administrative practitioner. Here again, the vision of how power was properly exercised rested on quantitative constructions of social realities. 'Statistical measurements provided a vocabulary of social analysis and grids of categories into which bureaucracies [could] attempt to fix the continuous mobility, tensions and dynamics of society.'[15]

If all of these facets of power in the idea of the modern state were manifest early in its history, one of the more conspicuous innovations of the twentieth century was a stronger, bolder emphasis on the modern state as the engine of large-scale social and economic planning. The idea of the state as the agent of social transformation was not wholly new. Baumann points out that the modern state emerged with a sense of the malleability of society in the hands of the politically powerful; or, changing the metaphor, society was seen as an open lawn which the state, conceived of as a gardener, would landscape and nurture. Equipped with the universalising standards of 'reason' and the means to measure its progress, the modern state represented

> popular, locally administered ways of life . . . as retrograde and backward-looking, a residue of a different social order to be left behind; as imperfect, immature stages in an overall line of development towards a 'true' and universal way of life, exemplified by the hegemonic elite . . . Those [retrograde] aspects were now seen as distinguished by their plasticity, temporariness, transitoriness – and, above all, amenability to purposeful regulation.[16]

This in turn became grounds for an ambitious confidence in the visionary powers of the state as the primary agent of social improvement. The twentieth century gave new form and vigour to these ideas, producing a burgeoning interest in Europe in the activity of national planning.

The first 'truly spectacular advance of national planning'[17] occurred in the socialist countries in which the state concerned itself with the planned development of the entire economy, not merely individual sectors or regions. With the mounting influence of Keynesian economic theory, spurred by the Great Depression of the 1930s, the idea of large-scale social and economic planning gained a wider currency in the liberal democracies of Europe and America, as an attempt to correct the perceived failures of the capitalist market. In Nazi

Germany the state espoused even greater confidence in its capacities to re-engineer the society in the image of an all-encompassing national plan.[18] In these emboldened versions of the 'gardening state', the rituals of scientific measurement offered a correspondingly more prominent contribution to the exercise of power. As the scale of the planning enterprise expanded, so the reliance on quantifiable measures of its object intensified. At the same time the use of statistics grew more elaborate. Large-scale social surveys – more sophisticated and wide-ranging versions of the research technique developed by European social reformers during the nineteenth century – harnessed the techniques of inferential statistics to generate ever more complex modes of measurement.[19]

Measurement and power pre-apartheid

If quantitative measurement of the population produced knowledge in a form well-suited to techniques of political control in Western states, it was likewise one of the epistemological underpinnings of power in many colonial states, where censuses and surveys were as much exercises in defining subject populations as in measuring their various demographic characteristics.[20] In the South African case, the creation of a unified state in 1910 launched a process of centralising and streamlining the exercise of power which would persist under various guises for at least another seven decades. At the outset, calls to integrate the country's four provinces through the enactment of 'uniform laws applying to the whole Union'[21] coexisted with strong and assertive traditions of the decentralisation of power to local and regional authorities (both rural and urban). From this perspective, the exercise of control through carefully crafted, often highly personalised, networks of paternalism and patronage was ideologically and strategically preferable to the impersonal rationality of modern bureaucracies. As one of the Under-Secretaries of Native Affairs put it, 'government officials must constantly bear in mind that... a full appreciation of local circumstances and conditions is the best way to secure goodwill, and that uniformity, however it may simplify office routine, is apt to be as great a danger as it is an advantage'.[22]

One of the themes which runs throughout the history of the South African state during the segregationist period is a continuing, if uneven, seepage of more modern visions and ambitions of power into the official mind. Some of the rituals of modern governance made their mark from the start. It was taken for granted that the bid for greater legislative and administrative 'uniformity'[23] presupposed nationally integrated measures of salient features of the population. So, along with the Act of Union came the Census Act, also in 1910, which provided for the first national census, duly conducted in 1911.²⁴ The Statistics Act of 1914 made provision for 'a complete system of statistics', entrusted to the Office of Census and Statistics, established in 1917.²⁵

Another of the devices of modern statehood enthusiastically adopted by the newly unified country was the official commission of inquiry. As Adam Ashforth argues, commissions of inquiry – with their aura of objectivity, independence and expertise – have played a central role in constituting 'a realm of discourse in terms of which the knowledge necessary for power can be discovered and expressed'.²⁶ 'Facts' are gathered, collated and presented, in the pursuit of 'solutions' to the 'problems' of power. But, if commissions of inquiry are instruments of modern statecraft, their versions of modernity are not static. So, too, the epistemological underpinnings of South African commissions of inquiry evolved through processes which tended increasingly to privilege 'scientific' notions of evidence – 'hard' measurable data – over the testimony of personal experience and political opinion. Research by Brahm Fleisch and Sue Krige into the making of Native Education policy, for example, suggests the increasing prominence of statistical notions of evidence within official commissions of inquiry from the 1930s.²⁷ The 1932 Carnegie Commission was the first commission to use large-scale surveys and statistical analysis as the basis of its findings.²⁸ But this newly 'scientific', quantitative notion of evidence was not yet routinised or institutionalised.

Ernst Gideon Malherbe played a crucial role in promoting and legitimising the use of social survey research more widely within state institutions. The National Bureau of Educational and Social Research, established by Malherbe in 1929, was the first initiative within the state to promote and institutionalise the contribution of quantitative social science to the processes of social reform. Lodged within the Department of Education and specialising in applied social research, the Bureau operated on a familiarly modernist assumption that scientific knowledge was the essential basis for rational, objective governance. As the then Minister of Native Affairs, Education and Social Welfare, H. Fagan, put it in 1939:

> the mingling of many races in South Africa and the great variety of physical and economic conditions may create many difficulties for the politicians, but at the same time they provide a wonderful field for those who wish to approach our problems by the method of scientific research. It is only recently that the importance and value of this method of approach has met with fairly general recognition. I feel strongly that the more our problems are tackled in this way, the more we will find that

their apparent complexity is no great cause for concern, and that they need not give rise to animosities and prejudices, but are capable of being dealt with in a practical and businesslike way.[29]

From Malherbe's standpoint, 'practical and businesslike' statesmanship was in turn integrally connected to the techniques of social measurement. 'The social and economic system', he argued, 'is in large measure subject to deliberate and scientific control',[30] which derived from its measurability. Reiterating the stance taken by the American psychologist E. L. Thorndike, Malherbe worked from the assumption that 'whatever exists at all exists in some amount [and] anything that exists in amount can be measured'.[31] Measurement could then provide the social planner with 'facts which have a definiteness and precision', sufficiently powerful to disrupt political complacencies and prejudices, and substitute an 'objective', scientific basis for social engineering.[32]

If the growing enthusiasm for 'scientific' statecraft was one avenue for the political power of social measurement, another derived from inherently quantitative definitions of some of the problems of social control and the manner of their solution by the state. This was nowhere more evident than in efforts to solve the so-called 'urban Native problem' – an 'excessive' exodus of African people from the reserves and rural areas of the country into the cities. Traditions of detailed counting in the process of labour recruitment, already well-established in the mining industry, were incorporated into the logic of the first national influx control policy, enacted in the 1923 Natives (Urban Areas) Act. Efforts to solve 'the urban problem' were seen to depend in large measure on finding formulae for limiting the numbers of African people permitted into urban areas according to the size of local (white) labour demands. The goal was to engineer, as far as possible, a three-way numerical match, between the numbers of African people permitted to be in an urban area, the numbers in employment and the numbers allocated township housing.

The substance of the 1923 Act was regularly disputed, within and beyond the Department of Native Affairs, and various local authorities managed their local 'Native problems' their own way. But the principle underpinning the 1923 Act continued to dominate the elaboration of state controls in the urban areas – as though the key to unlocking 'the urban Native problem' was in important respects a matter of arithmetic. To this extent, the pursuit of effective control was understood to depend on effective counting. Efforts to intensify the exercise of influx control were therefore closely associated with attempts to improve the frequency and volume of statistical counts of the size of the urban African population in relation to the size of the local African workforce. The 1937 Native Laws Amendment Act was intended to produce much stricter and more uniform influx controls, by empowering the Minister of Native Affairs to remove Africans from an urban area if they were 'surplus' to the 'reasonable labour requirements' of the area. To keep track of the size of the 'surplus', local authorities were compelled to take a biennial census of the resident African population in their areas of jurisdiction. In towns where this population exceeded 15,000 (which would have included all those clustered around sites of industrial development), the local authorities also had to provide the Department of Native Affairs with monthly returns showing 'the current employment position' and 'likely labour requirements'.

Yet, if the production and storage of social quanta was becoming increasingly routinised within the segregationist state, the project was as yet piecemeal and modest in aspiration, yielding few, if any, of the copious efforts at totalising measurement which would proliferate after 1948. These limits were particularly significant in respect of the measurement of 'race'. The Census Act of 1910 had provided for two types of census to be held at the discretion of parliament. The first type was to be carried out every five years, enumerating adult white males only 'for Constitutional purposes'; the second was an 'elaborate census' enumerating the entire national population every ten years.[33] In practice, limited censuses of 'non-whites' were undertaken in 1911 and 1946. But only one census 'covering all races' was conducted, in 1936. Another one had been planned for 1941, but was limited to whites only because of the war. By 1948, then, much to the chagrin of Afrikaner demographers, 'statistics with respect to the Bantu population of the Union [were] minimal'.[34]

Efforts to produce a calculus of African labour supply and demand were even more desultory. Although the intention of the 1923 Natives (Urban Areas) Act had been to create a nationally uniform system of influx control, the terms of the law allowed local authorities to opt in or out of its provisions. Most stayed outside, which meant that the calls for statistical measurement and regulation of the African labour market were widely ignored. The 1937 Native Laws Amendment Act attempted to introduce greater uniformity within the sphere of urban native administration, but the Department of Native Affairs was deeply divided over its merits and failed to oversee its implementation in any consistent way.

World War Two represented a watershed in problematising the practice of government, in ways which in turn produced calls for more centralised, comprehensive modes of the measurement of population.

The 'Native problem' was at the heart of this discursive shift. As vast numbers of African people entered the cities, the state's existing capacities for control in urban areas were completely overwhelmed. Most local authorities were wholly unable to stem the tide of African migration.[35] At the same time housing construction in the townships came to a virtual standstill, producing overcrowding, accommodation shortages and explosions of popular discontent on a scale hitherto unknown. Squatter settlements in areas beyond municipal control also proliferated in abundance – perhaps the most powerful signifier for white society of chaos, the absence of effective control, and the threat of moral and political menace.

One of the decisive effects of the war was to provoke heated re-evaluations of statecraft within the state and white polity at large, foregrounding the issue of managing the 'problem' of race. It was initially the local authorities which complained most volubly that the Department of Native Affairs was ill-equipped to tackle the problems of control looming in the urban areas. The main weakness they identified was the absence of rational planning: 'attempts at proper planning came into the picture only when the problem within a specific area became accentuated out of all proportion to its solution'.[36] Local authorities which had themselves been firm supporters of decentralised administration now saw this arrangement as something of an anachronism and called for more integrated, centrally planned and coordinated interventions from the Department of Native Affairs.

The Department of Native Affairs remained lethargic and indecisive. But calls to extend and modernise the apparatuses of urban control – and the state more broadly – began to resonate more widely. The establishment of the Council for Scientific and Industrial Research (CSIR) in 1946 was in part an effort to institutionalise the contribution of 'scientific' expertise to processes of government, as befitted a 'modern' state. The Social and Economic Planning Council (SEPC), instituted in 1942, was borne of what Peter Wilkinson has rightly called a 'discourse of modernity', distinguished by its emphasis on coordinated, 'scientific', state-led planning of the country's social and economic development.[37]

The thinking of the SEPC anticipated, in some respects, the mode of political rationality that informed the Fagan Commission of 1946 (which subsequently fed into the governing party's 1948 election manifesto). Although uncertain and ambiguous at times, Fagan echoed local authorities in recommending that 'the urban problem' had grown far too large for small-scale, locally uneven, *ad hoc* interventions. To solve a 'problem' of such proportions necessitated a more centralised, 'well-planned', nationwide assault, mounted within a more comprehensive and sophisticated network of state institutions. This included, in particular, the creation of a national system of labour bureaux to monitor and control the flow of African people in and out of the cities, regulated by the pulse of local labour markets. Fagan stressed that this in turn required the development of effective systems for identifying and classifying African people so as to 'ensure that everybody has some fit place to which he is entitled to go'.[38] Fagan also recommended the expansion of central state bureaucracies and redirection of powers away from the local authorities. In his view, control was to be effected through the rational, efficient state mastery over impersonal demographic processes. Undergirding his recommendations was the modernist assumption that states needed to get bigger at the centre, in order to broaden the ambit of their controls. If they did so appropriately, there was no problem which defied solution.

On the eve of the 1948 election, the broad thrust of Fagan's call for a stronger, more centralised state, was enthusiastically supported by the National Party as well. The Afrikaner nationalist alliance was deeply divided over the meaning of apartheid as well as the means of achieving its objectives. The two different renditions of apartheid – 'total segregation' versus a more 'practical' project – envisaged different orders of social transformation.[39] Nevertheless, they agreed on the necessity to plan for the preservation of white supremacy and white racial purity and that apartheid had to be more whole-hearted and systematic than segregation had been. Both tendencies treated apartheid as a project of national proportions which could only be undertaken by a bigger, stronger and more knowledgeable state and with Afrikaners at the helm.[40] Taken in conjunction with Fagan's recommendations, we can safely assume that a more self-consciously 'modern' state was likely to result whatever the outcome of the 1948 election.

The transition to apartheid

What then, made apartheid different, both from the segregationism which preceded it and from other regimes of power elsewhere in the world? Recently, Mahmood Mamdani has railed against the sense of South Africa's exceptionalism. He presents an alternative view of apartheid as 'the generic form of the colonial state in Africa', defined as the 'bifurcated' unity of two different modes of governance, direct and indirect rule.[41] Mamdani's intervention serves a useful purpose in its insistence on contextualising South Africa within the rest of Africa and in framing questions about the logic of rule in ways which do not deflect the issue to the matrix of interests underpinning it. We should go further though, in recognising the resonances between the political logic of apartheid and more global adventures in modernity. As the first section of the paper suggests, the Nationalists' confidence in the enormous problem-solving capacities of a large, efficient, well-informed state echoed some of the ideas about modern statecraft which prevailed in many other parts of the world at the time. Roosevelt's New Deal, the British state's programme of post-war reconstruction, or the Nazi programme for Aryan supremacy, were each particular instances of a wider consensus about the prospects for large-scale social transformation, and they all evinced a profound confidence in a suitably 'modern' state's capacities to bring it about.[42]

Yet, if South Africa's exceptionalism should not be overstated, nor should the issue of its distinctiveness be ignored. If Mamdani's analysis fails to address the question, the most common answer to it in the South African literature has been to see apartheid as a 'more systematic', and therefore 'more intense', system of racial exploitation than its segregationist predecessor – an answer which typically fails to specify what 'more systematic' actually means, and whether this produced a difference merely of degree, or a difference in kind. As I see it, apartheid produced a distinctive kind of system borne of a particular idea of systematicity in which modern ideas about the 'gardening state' promoted visions of the bureaucratisation and surveillance of race in unique and unprecedented sorts of ways. If, prior to 1948, aspects of modern governmentality infused parts of state policy and its institutional fabric in a piecemeal sort of way, after 1948 the commitment was determined and thorough-going. Afrikaner nationalism's political morality of race and racial difference was grounded in religious discourses with strains of German idealism;[43] but when it came to the construction of power itself, apartheid represented a concerted bid to 'modernise' racial domination.[44] By re-imagining the capacities, scopes and entitlements of the state along more assertively 'modern' lines, it became possible in turn to imagine a society in which constructs of race became the all-embracing basis of the social order fashioned by the state, and the unit of what was envisaged as a ubiquitous, all-encompassing classification and quantification of the country's population. The apartheid version of the 'modern' state was one which was sufficiently large, powerful, bureaucratically expert and knowledgeable to keep each race in its proper place, economically, politically and socially.

The making of apartheid was therefore rooted, first and foremost, in a rethinking of the idea of political power. Since World War Two, the Nationalists argued, South African society had become rampantly chaotic and disorderly – an anxiety captured most vividly in the image of the racial 'oorstroming' (overflowing) of the cities. Black people, they insisted, were not simply overrunning the numbers of whites in urban areas, they were jeopardising the entire racial social order. 'Problems' of this magnitude could only be solved through the intervention of the state on a national scale, leaving no stone unturned – which in turn necessitated a much enlarged, centrally unified state exercising comprehensive powers on the basis of careful planning. Apartheid statecraft represented a hankering for 'total order' – although at no stage did its architects succeed in producing a single 'grand plan' in the image of which to pursue it.

The central state duly grew much bigger and its ambitions expanded accordingly.[45] The architects of apartheid had high expectations of the rationality of government as a coordinated response to the problems of the day. Within the Department of Native Affairs, the discourse of planning became newly salient. The call for 'proper planning' became one of the mantras of 'native policy', as did the attempt to 'create order from chaos',[46] to eliminate the 'laissez-faire policy of the past decades'[47] and to rationalise the use of African labour in 'efficient', 'orderly' ways.[48]

Hendrik Verwoerd was one of the most assertive champions of the modernist logic of power. His faith in the socially transformative powers of science, as the basis for informed and rational planning, was nurtured during his years as an academic with a particular interest in 'the poor white question'.[49] In 1934, Verwoerd reviewed Roosevelt's New Deal, and applauded his 'willingness to reshape the state extensively in the pursuit of far-reaching goals defined by the state itself'.[50] Verwoerd's modernist bent was also evident in his support for moves among urban 'native administrators' themselves to 'improve the professional and technical knowledge of Administrators',[51] creating a specialist qualification in 'Non-European administration' which took cognisance of its 'scientific aspect'.[52]

These attempts to produce new technologies of rule included aspirations for newly systematic modes of measurement. After all, racial 'oorstroming' was fundamentally a quantitative issue – a central component of 'the political arithmetic of the South African population', as Afrikaner demographer J. L. Sadie put it:

> The population, its growth or decline, births and deaths, the racial composition, are the basic data of politics ... In South Africa the outstanding problem, dominating all others, is the relative numbers of the different races constituting the Union's population, and their differential rates of growth.[53]

Little time was therefore wasted in changing the form of the quinquennial national census to include all races – in time for the first one in 1951. The Nationalists also instituted the process of creating the first national register of births and deaths for the African population.

But apartheid's 'political arithmetic' went much further. Efforts to produce nationally complete measurements of the different racial groups and their relative rates of growth presupposed a uniform, comprehensive system of racial classification. The Population Registration Act, promulgated in 1950, was intended to provide the basis of a totalising system of racialised measurement and surveillance. The National Population Register proposed by this new law was imagined as the basis of an integrated, authoritative system of information linking an individual's racial classification to other legal and administrative domains of knowledge. Such a unified data base, the Minister of Interior suggested, would 'be of great assistance in the administration of various industrial and social welfare laws... furthermore it will be of great importance in controlling unlawful immigration and crime [and] can also be of great use with regard to the administration of Native legislation and in the application of the principle of apartheid'.[54]

To a much greater extent than preceding regimes, apartheid's policymakers tried to comprehend the country as a whole, rather than as regional segments with particular, often distinct, political and administrative traditions. In most instances, this was seen to depend on being able to aggregate each local manifestation of the various 'problems of control' – from unemployment, squatting and juvenile delinquency through to inter-racial sex – on a countrywide basis. For Verwoerd in particular, comprehending each problem as a 'whole' was the hallmark of a 'scientific' approach to its solution.[55] The exercise was inherently quantitative: measuring each instance of a problem and then adding them up was the only way to represent the national dimensions of the problem.

The links between counting and controlling were most elaborate in respect of the African population, particularly in the urban areas. The effort to solve 'the urban native problem' was a multi-pronged attempt to regulate the numbers of Africans legally entitled to live and work in a prescribed urban area. Section 10 1 (a) (b) (c) of the Urban Areas Act (as amended in 1952) specified three sets of conditions under which Africans could live in an urban area irrespective of whether or not they were employed there. For the rest, the right to live in an urban area depended on securing authorised employment through the offices of the labour bureaux system. The labour bureaux system was designed to operate according to an urban labour preference policy. Having assiduously courted the votes of white farmers, the Nationalists were careful to stress that the impact of 'the urban problem' was felt far beyond the cities in rural areas all over the country. If the problem in the cities was one of 'oorstroming', many of the country's white farmers complained of acute labour shortages as hundreds of thousands of farm workers fled to the cities. The 'urban Native problem' then, was the obverse of a rural 'Native labour problem', which had to be treated as a unified national problem, rather than a series of local problems of control. Countrywide, said Secretary for Native Affairs W. M. Eiselen, African labour was 'maldistributed'. Solving the problem required taking an integrated view of the country's labour markets, both urban and rural, so as to engineer a more effective, 'rational' allocation of labour between them.[56] 'The whole modern labour system', said Verwoerd, 'is one of planning, one of not allowing people to roam aimlessly until at some time or other a place is accidentally found where there are possibilities of employment.'[57] This, said Verwoerd, was the 'modern' way – suitably adapted to the South African situation.

The plan which duly emerged – an urban labour preference policy – was a remarkable demonstration of the hubris of its architects. The new cadre of Nationalist leadership was confident of the state's powers to control market forces in what they deemed 'more rational' ways. No problem – not even the political refashioning of the market – was too large or overwhelming to defeat rational, efficient efforts by the state to solve it, in time. And the basis of this extraordinary confidence was another exercise in what Sadie called 'political arithmetic', this time by way of a simple quantitative model of how to engineer 'rationality' in the country's labour markets. Labour supply and demand were both conceived as homogeneous quanta, which ought simply to be numerically matched. So, if in any given city, the collective demand was for x number of African workers and the size of the local economically active African population was $x + n$, then it was deemed irrational to bring more labour into the area, until the growth in the size of the local labour demand exceeded n.

This strategy for restoring 'order' to the urban areas presupposed the routine production of a range of statistical data. What the Department of Native Affairs had in mind was a horizontally layered, vertically integrated system of quantitative information-gathering, with local authorities compiling masses of local data, more composite pictures being produced at the district and regional level, and the central Department synthesising and collating all of these measurements with panoptic completeness.

During the 1950s the demands on the counting capacities of the urban local authorities grew to be enormous. In order to keep track of the size of the local African population, various sources of data needed to be compiled and updated on a regular basis – all done laboriously by hand since local authorities could not afford any mechanical equipment. The first was a series of records maintained by the superintendent of each and every township within the local authority's area of jurisdiction.

> Every Superintendent of a Bantu residential area is required to maintain 'a house file' in respect of every family accommodated in his Township in addition to the ordinary rent record cards. When a house is initially allocated to a family, the tenant is interrogated and a form is completed setting out the basic personal particulars of the registered tenant, his wife, the children, any relatives or dependents living with that family, as well as all sub-tenants and temporary visitors. In compiling this information an attempt is made to record the ages and sex of each person involved, the date of first entry into Johannesburg, and as much information as is possible in regard to employment, and the influx category into which the person falls.[58]

Similarly detailed data was required in respect of every resident of all the hostels. The local authority was also expected to produce aggregate counts on the basis of all of these records, and then check these figures against the results of the national population census in these areas.

Another data base, compiled by a different office, comprised labour records for all African males employed in the area, repeating much of the data assembled by the Superintendents but for the purpose of keeping detailed counts of monthly ebbs and flows in local labour supply and demand. Meanwhile, other offices were expected to keep tabs on employers' payments to register their African workers (to be cross-checked with the figures for numbers of workers employed). And copious statistics were required in respect of Africans living and working on all 'licenced and unlicenced premises' in the city.

The complexity and volume of these statistical instructions increased along with shifts in 'Native' policy. The 1960s saw the advent of a new order of social engineering, accompanied by even more bluster about the virtues of 'proper planning'. In the early part of the decade, new-found efforts to contain the size of the urban African population produced a flurry of meetings, committees and draft legislation aimed at eliminating perceived anomalies in existing policies, removing any vestiges of decision-making power exercised by local authorities, and subordinating labour markets to newly aggressive modes of state intervention. Plans proliferated for large-scale population removals, as well as remodelled strategies for 'canalising' surpluses of urban labour to areas of shortage on a regular basis, along with the imposition of labour quotas in officially designated areas. These new policy initiatives were accompanied by a discursive renaming of the African population. In a bid to celebrate the essential 'tribal' unity and integrity of all Africans, rural- and city-dwellers alike, the Department of Native Affairs now recategorised the African population as a series of officially designated 'ethnic groups'.

These discursive and policy shifts issued in copious instructions for even more frenzied efforts at measurement. At the township level, local authorities were now required to compile all their regular data according to ethnic group. Moreover, by 1963, in line with the more aggressive approach to influx control, the grid of legal and administrative controls over residence and employment which was previously applied in respect of African men only, now became fully applicable to African women too, which meant that gender breakdowns were also added to each aspect of the statistical work of the local, district and regional labour bureaux.

Things became even more complicated from the mid-1960s as the state restated its commitment to 'orderly planning' with some new innovations. Legislation in 1965 provided for the creation of a new Department of Planning in order to enhance the planning process. The ambit of this higher order of planning included new efforts at 'physical planning', in which efforts to redraw the map of the country's industries were yoked to new dimensions of the political arithmetic of population. The 1967 Physical Planning Act formalised a strategy which the Department of Bantu Administration and Development (formerly the Department of Native Affairs) had been promoting since 1960 to contain the numbers of Africans in urban areas by disallowing the siting of new industries in these areas unless their ratio of African to white workers conformed to official specifications. One of the reasons for the delay in promulgating this legislation was the Department's difficulty in producing what it considered the appropriate numerical formula. Officials finally settled on a ratio of 1:2.2 as the maximum permissible ratio of white to African workers in so-called 'white' industrial areas.[59]

Although the newly created Department of Planning was responsible for administering this legislation, the labour bureaux were required to produce the relevant statistical data in respect of each and every industrial establishment in their area, showing 'the number of Bantu presently employed' as well as 'the maximum number of permissible Bantu employees'.[60] By measuring the size of labour forces per industrial establishment, labour bureaux could then supposedly make decisions about the allocation of new work-seekers' permits on the basis of a calculation of the maximum number of workers which any establishment could sustain without contravening the specified maximum ratio of white to African workers there.

Another policy innovation of the late 1960s (also first mooted much earlier on in the decade) was the programme of large-scale population removals, during which the state sought to eradicate 'black spots' from 'white areas' as well as 'resettle' all individuals living in urban areas who were considered 'unproductive' and therefore part of an urban population 'surplus'. All resettlements were to be on the basis of ethnic homogeneity. The logic of this policy also depended on rituals of extensive counting – in this case, measures of the numbers of people in black spots, together with the numbers of 'unproductives', all broken down according to ethnic group. Local authorities were also required to assemble additional data in respect of 'non-productive Bantu': officials had to count these people according to the categories of 'registered female tenants', 'registered female lodgers', 'registered professional Bantu', 'registered industrialists and traders', and then supply between ten and fourteen pieces of information in respect of each one.[61]

By the end of the 1960s, the scope of the Department of Bantu Administration and Development's obsession with measurement was nothing short of breathtaking. By now a huge and labyrinthine department with many divisions and subdivisions, its own routine statistical enterprise included: monthly counts of the numbers of work-seekers' permits issued on a national basis; the number of service contracts registered (showing the area of origin of each of the registered workers); the numbers of registered vacancies in urban areas and in bantustans and on farms; annual counts of the numbers of 'Bantu' seeking assistance from special 'aid centres' which assisted with the work of the labour bureaux; the compilation of an 'occupation register... in respect of each national unit... containing certain particulars of the Bantu concerned'; the creation of a separate data base of 'existing and anticipated employment opportunities in the homelands for Bantu with advanced qualifications'; annual counts of the number of workers recruited by various recruiting organisations and the sites of recruitment; annual counts of the numbers of curfew proclamations issued, the numbers of 'new Bantu residential regulations' promulgated as well as amendments to the existing regulations; annual counts of the number of bodies designated as urban local authorities and the number of 'promulgations and redefinitions of Bantu residential areas'; annual counts of the numbers of permits issued in terms of the Group Areas Act for 'recreation and health services', such as cinemas and hospitals; numbers of applications to 'conduct church services for Bantu

A MANIA FOR MEASUREMENT

in white residential areas'; amounts of money accruing in respect of the Bantu Services Levy Act; annual income from the sale of 'Bantu Beer' by local authorities and employers; numbers of inspections undertaken of a whole range of different types of sites on which particular types of developments were being considered (each type of site being enumerated separately); 'numbers of cases in which comments ... from inspectors were furnished on group areas planning'; numbers of townships planned and developed, per ethnic group; numbers of sub-economic houses constructed; numbers of families 'removed' from urban townships, on an ethnic basis; numbers of families removed from 'black spots' and white rural areas, per ethnic group; separate counts of numbers of 'Bantu traders, industrialists and professionals' resettled; numbers of children placed in various welfare institutions; annual counts of numbers of reference books issued to males and females; numbers of duplicate reference books issued to males and to females; numbers of identification documents issued to 'foreign Bantu'; comparisons of 'numbers of Bantu whose identity numbers were known with numbers of existing records of fingerprints'; numbers of fingerprints 'classified and searched to determine whether the fingerprints of the persons concerned are not already on record'; numbers of sets of fingerprints 'added to existing record'; numbers of births, marriages and deaths registered each year; annual counts of the numbers of 'Bantu males' on the National Population Register; annual counts of the numbers of 'Bantu females' on the Population Register; numbers of enquiries into 'the tax particulars and movements of Bantu' on the Population Register; numbers of beneficiaries of social pensions, maintenance grant beneficiaries and pneumoconiosis grant beneficiaries; along with records of monies spent on salaries, capital equipment, land and other projects.[62]

The Department of Bantu Education (previously a part of the Department of Native Affairs) set to work with similarly demented attention to detail. Announcing a more intense devotion to the activity of 'planning',[63] the Department's annual report for 1965 was replete with large tables spread over several pages, showing *inter alia*, teachers' remuneration according to the type of school in every region, broken down by gender and race; the distribution of teachers according to mother tongue or ethnic group, broken down by race, gender, and in the case of the 'Bantu' also by ethnic group, and type of school (e.g. private, government, community, mine, hospital, Catholic church schools, non-Catholic church schools) and the level of instruction (primary, secondary and technical); the classification of teachers according to type of school, level of instruction, gender and age cohort.

The fervour for more and more figures during the mid- to late 1960s made its mark in the central Bureau of Census and Statistics too. In 1965, the Statistics Act was amended to allow for a huge expansion in the scope of statistical data collection into 'social matters and activities of whatever nature, family and household surveys including surveys of family and household budgets and any other matter prescribed by the Minister'[64] – areas of experience previously considered 'private' and therefore protected from statistical surveillance.

In short, by importing the quest for thorough-going racialisation into the modernist project of government, the apartheid state produced its own unique sort of monster – a statistical frenzy which extended well beyond the routines of measurement in most modern states.

Implications for the exercise of power

Apartheid was never implemented wholly as intended. The designs of some policies were internally contradictory; others were at odds with each other. Many of the ambitions of apartheid's planners were never fully realised, and, often, control was established in different ways, and to different degrees, from those intended. The exercise of power was always contested, in ways which profoundly affected the practice of a policy, as compared with its original intentions. Yet even so, the state did achieve an unprecedented degree of centralised control and it invaded the lives of its citizens in much more labyrinthine and aggressive ways than before.

These more stringent controls were rooted in more copious counting. For those African people unlucky enough to have been enumerated in one or other way, surveillance of their activities grew increasingly acute as apartheid developed. If the local labour bureau knew the full record of a registered work-seeker, then that person probably had a relatively difficult time eluding detection by trying to slip into the urban area once his/her permit expired without him/her having entered into authorised employment. The removal of over three and a half million people to officially designated tracts of barren land during the late 1960s and early 1970s could not have been undertaken without the measurements of population and property produced by the Department of Bantu Administration and Development – even if much of the detail was superfluous.

The complementarity of counting and controlling was also less directly instrumental. The state exercised some power in the mere fact of being able to impose and routinise activities of counting. The rituals of statistical measurement became a mode of discipline within the state as much as a technique of surveillance within the society at large:

A MANIA FOR MEASUREMENT

those who did the counting, as well as those who were counted, knew that they were being watched. The exercise of power was also effected through the institutionalisation of a racialised statistical discourse. The ubiquity of racialised classification and counting contributed enormously to normalising the state's racial grid, which was arguably the bedrock on which the routines of apartheid rested. As Ian Hacking puts it, 'the bureaucracy of statistics imposes not just by creating administrative rulings, but by determining classifications within which people must think of themselves and of the actions that are open to them'.[65] Under apartheid, the elaborate rituals of racialised measurement constructed the population as discrete, racial units, and the apparent objectivity of this statistical discourse gave political manoeuvres the respectability and rationality of modern 'science'.

As the reliance on statistical measurement grew increasingly routinised, so too did confidence in the idea of a statistical 'fact' as an authoritative measure of social reality. Statistics became part of the rhetoric of authority and expertise.[66] Yet, in other respects, the apartheid state's aspirations for a totalising mode of control amounted to rather less in practice. There are many factors which account for this, among them, the limits of the state's statistical 'gaze'. This was partly a product of a recurring and increasingly serious shortage of statistical capacity and competence within the state, at all levels and sites of its statistical enterprise – a problem which worsened with the escalating demands for statistical surveillance. In 1951 the Director of the Bureau of Census and Statistics had already identified 'a shortage of capable staff, particularly statistically qualified staff', together with a large number and high turnover of temporary staff.[67] By 1969, the dearth of statistical expertise had become seriously disabling, preventing the Bureau from fulfilling aspects of its brief.[68]

A paucity of statistical expertise within the civil service at large might account for the amateurish ways in which several departments presented and compiled their statistical data, typically as raw data, with little by way of statistical interpretation.[69] Limited skill might also account for conceptual muddles over what was being measured. Data compiled by individual departments often omitted to specify the meanings of the statistical categories used, which made many of their tables opaque and unreliable.[70]

Unlike any other departmental annual reports, those from the Department of Bantu Administration and Development conscientiously avoided any discussion of the issue of staff shortages, but there are many other indications that here, too, the reality of the statistical enterprise fell far short of the ludicrous expectations of its engineers. The production of labour statistics – one of the Department's most laborious undertakings – was punctured by the continuing administrative failures of the labour bureaux. By 1967, despite repeated attempts to improve and strengthen the system, the Department noted that many urban areas had still not established labour bureaux, and in those areas where labour bureaux did exist, there was nevertheless 'wholesale confusion' among bureaux officials as to how to put their manifold administrative instructions and bureaucratic procedures into practice. 'The prescribed cards and records were either simply not kept or were incomplete'[71] – hardly surprising given the paucity of bureaucratic expertise on one hand, and the preposterous enormity of the Department's instructions on the other.

Even those piles of statistics which were compiled had limited utility (other than as quantitative indices of the labour bureaux' performance). The purpose of such data-gathering was to allow for the operationalisation of the urban labour preference policy, in terms of which no work-seeker from outside an urban area would be allowed in if there was a 'surplus' of local labour in the area; hence the need to measure local labour supply, demand, and keep track of the numbers of work-seeking permits issued to outsiders. But in most urban areas these figures had little bearing on the day-to-day routines of influx control because in practice, labour bureaux officials generally did not make decisions about whether or not to issue more work-seekers' permits or register service contracts in the light of statistical measurements of the local labour supply – in the knowledge that the Department's arithmetical model of the labour market made absolutely no practical economic sense.[72]

Given the arcane details with which many sources of data were compiled, the fate of labour bureaux statistics was one instance of a broader pattern: many statistical minutiae were assembled for inclusion into annual reports – such as the numbers of permits issued for 'church services for Bantu in white residential areas' or the numbers of Zulu-speaking female teachers between the ages of thirty and thirty-nine in non-Roman Catholic church primary schools – but with little, if any, practical bearing on either the making or implementation of government policy.

Public reports of limited statistical expertise perhaps also account for the traditions of scepticism and suspicion which developed internally within the state between its various statistical agents. Once the National Party took office, its policy 'experts' soon announced their mistrust of the reliability and scope of the statistical data compiled by the previous regime.[73] But similar expressions of uncertainty lingered throughout the apartheid period. For example, when a departmental committee appointed by the Department of Bantu Administration and

Development set out to produce a series of new policy recommendations on influx control in 1962, it was altogether dismissive of local authorities' efforts at statistical measurement.[74] Simon Brand told me during an interview in 1984, soon after he had become the chairman of the newly created Development Bank, that one of the first and most important tasks of the Bank was to generate its own set of statistical data because he mistrusted all of the data produced to date by the Bureau of Statistics.

Whatever the merits or demerits of such blanket judgements, there are particular reasons to mistrust many of the huge heaps of statistical data accumulated in respect of the African population. Published and unpublished reports evaluating the implementation of apartheid policy routinely cited statistics of urban and rural African employment, unemployment and labour bureaux placements, alongside measurements of population size and population growth. But the mounting rigour of the state's policies towards Africans corrupted all of these counts, as many thousands of people with no legal 'right' to live in the urban areas eluded the authorities. For obvious reasons 'illegals' strove not to make their presence known, exhibiting what W. J. P. Carr, Manager of the Johannesburg Non-European Affairs Department, quaintly called 'the inborn reluctance of the Bantu people to be counted'.[75] Intensifying the drive to control the African population only compounded the problem. Stricter laws did not wholly deter people from coming to town; by shrinking the spaces for legal migration, they created a larger population of 'illegals'. And the larger the number of 'illegals', the harder it was to police laws effectively.

Statistical counts of the African population were thus also emblematic of the limits of surveillance. The practice of apartheid created endemic forms of ignorance for the state; to be 'illegal' was by definition to be invisible to official eyes. The more stringent the state's practice, the more entrenched the ignorance. So, when Verwoerd asserted (for example) that 'it is well-known that there are between six and eight hundred thousand illegal immigrants in the country',[76] or when the Johannesburg Non-European Affairs Department produced lengthy counts of the numbers of 'illegal Bantu' in Johannesburg[77] – the use of statistics confers an illusion of precision in professing to count the uncountable.[78]

Conclusion

In his article on Geoffrey Cronjé, J. M. Coetzee makes a plea for 'getting irrationality into history'[79] in order to take account of apartheid's collective madness. But, as Saul Dubow has warned, 'when discussing apartheid, it is important not to draw false distinctions between rationality and irrationality, sanity and madness: there was method in the madness of apartheid, just as there was madness in its method'.[80] This chapter has discussed apartheid as a mode of rule in which the hankering for thorough-going racialised 'order' produced an idea of state power as a demented sort of rationalism, borne of an obsession with the modernist logic of measurement.

Yet it would be wrong to present apartheid's modernist yearnings, intense and grandiose as they were, as having issued in an altogether comprehensive, unitary or coherent political project. This chapter has isolated some of those aspects of apartheid statecraft which fitted well with the idea of a more centrally interventionist state, honing its powers of surveillance through its capacities to construct comprehensive systems of racialised population measurement and classification. But the challenges of political control were by no means uniform and there was much that the Department was unable to 'see' or count. The Nationalists' pursuit of 'order' was ultimately limited as much by bureaucratic incompetence as by the recognition of – and capitulation to – the realities of the market which the state completely failed to refashion.

These lapses in the control conferred by the routines of measurement are not merely academic ironies in the practice of apartheid. They signal the place and impact of social agency in complicating, and at times confounding, the knowledge/power nexus in ways that often disoriented decision-makers, destabilised their plans, and subverted the prospects for the kind of integrated, systematic and all-encompassing modes of control to which the apartheid state aspired. Nicholas Thomas' remarks about the limits of colonial knowledge are equally apposite to the apartheid project:

> even if colonial knowledge often took the form of a panoptical, encyclopaedic appropriation of indigenous customs, histories, relics and statistics, such displays of intellectual rapacity were frequently accompanied by a kind of despair, which found the space and social entity of the colony to be intangible, imperceptible and constantly untrue to the representations that might be fashioned of it.[81]

Notes

1 An earlier version of this chapter was presented as the Keynote Address to the annual conference of the Canadian African Studies Association, in May 1995, and as a seminar in the Institute for Advanced Social Research and Department of Sociology, University of Witwatersrand, 1996.

A MANIA FOR MEASUREMENT

2 I. Hacking, 'How do we do the history of statistics', in G. Burchell et al. (eds), *The Foucault Effect* (Chicago, 1991).
3 *Ibid.*, p. 186.
4 *Ibid.*, p. 181.
5 S. Woolf, 'Statistics and the modern state', *Comparative Studies in Society and History*, 31:3 (1989), 603–4.
6 M. Foucault, 'Governmentality', in Burchell et al. (eds), *The Foucault Effect*, pp. 87–92.
7 C. Gordon, 'Governmental rationality: An introduction', in *ibid.*, p. 9.
8 Foucault, 'Governmentality', p. 99.
9 Gordon, 'Governmental rationality', p. 36.
10 Foucault claims that the conceptualisation of problems of 'population' which inheres in modern governmentality displaced earlier notions of the governance of 'peoples'. But in the South African case – and probably all colonial societies – these two modes of reasoning were by no means mutually exclusive; rather, they coexisted and intersected. See also A. Stoler, *Race and the Education of Desire* (Durham and London, 1995), p. 9.
11 Foucault, 'Governmentality', p. 99.
12 T. Johnson, 'Expertise and the state', in M. Gane and T. Johnson, *Foucault's New Domains* (London, 1993), p. 142.
13 Z. Baumann, *Intimations of Postmodernity* (London and New York, 1994 edn), p. 6.
14 M. Weber, 'Politics as a vocation', in H. Gerth and C. W. Mills, *From Max Weber* (London and Boston, 1977 edn).
15 Woolf, 'Statistics and the modern state', p. 604.
16 Baumann, *Intimations of Postmodernity*, pp. 7–8.
17 S. Gomulka, 'National economic planning', in W. Outhwaite and T. Bottomore (eds), *Blackwell Dictionary of Twentieth Century Social Thought* (Oxford and Cambridge, Mass., 1996 edn), p. 404.
18 Z. Baumann, *Modernity and the Holocaust* (Cambridge, 1995 edn).
19 P. Halfpenny, 'Social statistics', in Outhwaite and Bottomore, *Blackwell Dictionary of Twentieth Century Social Thought*, p. 610.
20 See e.g. B. Anderson, *Imagined Communities* (London, 1991 edn), chapter on 'Census, map, museum'; B. Cohn, *An Anthropologist among the Historians and Other Essays* (Delhi, Oxford and New York, 1992 edn), chapter on 'The census, social structure and objectification in South Asia'; N. Thomas, *Colonialism's Culture* (London, 1994).
21 Union of South Africa, *House of Assembly Debates* (henceforth *HAD*), 1910/11, col. 12.
22 Quoted in J. Allison, 'Some aspects of urban native administration', *Race Relations*, 7 (1940), 112.
23 *HAD*, 1910/11, col. 12. See also S. Dubow, *Racial Segregation and the Origins of Apartheid in South Africa, 1919–36* (London, 1989).
24 Traditions of regional census-taking were well-established prior to Union.
25 Union of South Africa, Bureau of Census and Statistics, *Official Year Book*, no. 27 of 1952–53, G.P.S. 12694, 1953–54.
26 A. Ashforth, *The Politics of Official Discourse in Twentieth Century South Africa* (Oxford, 1990), p. 5.
27 B. Fleisch, 'Social scientists as policy makers: E. G. Malherbe and the National Bureau for Educational and Social Research, 1929–1943', *Journal of Southern African Studies*, 21:3 (1995); S. Krige, 'Segregation, science and commissions of enquiry: The contestation over native education policy in South Africa, 1930–1936', *Journal of Southern African Studies*, 23:3 (1997), 491–506.
28 Fleisch, 'Social scientists as policy-makers'.
29 H. Fagan 'Foreword', in E. G. Malherbe, *Educational and Social Research in South Africa*, South African Council for Educational and Social Research Series, no. 6 (Pretoria, 1939).

30 Malherbe, *Educational and Social Research in South Africa*, p. 7.
31 Cited in *ibid.*, pp. 28–9.
32 *Ibid.*, p. 38. See also S. Dubow, *Scientific Racism in Modern South Africa* (Cambridge, 1995), p. 215.
33 *HAD*, 1910/11, cols 61–2.
34 J. L. Sadie, 'Some notes on Bantu demography', *Journal of Racial Affairs*, 2 (1955), 36.
35 See D. Posel, *The Making of Apartheid, 1948–1961* (Oxford and Johannesburg, 1991 and 1997), ch. 2.
36 A. S. Marais (Manager for Non-European Affairs, Boksburg), cited in Institute for Administrators of Non-European Affairs (hereafter IANA), 'Record of proceedings of first annual conference, 1952'.
37 P. Wilkinson, 'A discourse of modernity: The Social and Economic Planning Council's fifth report on "regional and town planning", 1944' (mimeo, n.d.), pp. 7–8.
38 Union of South Africa, *Report of the Native Laws Commission, 1946–8*, UG 30/1953, p. 27.
39 Posel, *The Making of Apartheid*, chs 1 and 2. See also J. Lazar, 'Conformity and conflict: Afrikaner nationalist politics, 1948–1961', D.Phil. thesis, University of Oxford, 1987.
40 But this is not to suggest the absence of division or contestation over the desiderata or limits of the process of 'modernity'. Tim Clynick has argued that continuing conflicts within the ranks of the Afrikaner nationalist alliance throughout the 1950s reveal competing 'manuscripts for modernity'. See T. Clynick, 'A boerevolk for a boerestand?: Broedertwis and two manuscripts of platteland modernity, c. 1959', Seminar Paper no. 367, Institute for Advanced Social Research, University of Witwatersrand, 1994.
41 M. Mamdani, *Citizen and Subject* (Kampala, Cape Town and London, 1996), p. 8.
42 See e.g. Baumann, *Modernity and the Holocaust*.
43 T. Dunbar Moodie, *The Rise of Afrikanerdom: Power, apartheid and the Afrikaner civil religion* (London, Berkeley and Los Angeles, 1975).
44 Heribert Adam used the phrase several years ago, but its applicability was limited to processes of 'pragmatic rationalisation' – an instrumental reasoning – which characterised decision-making within the apartheid state, particularly in the later years. See H. Adam, *Modernizing Racial Domination* (Berkeley, Los Angeles and London, 1971).
45 D. Posel, 'Whiteness and power in the South African civil service: Paradoxes of the South African state, *Journal of Southern African Studies*, 25:1 (1999), 102–7.
46 SABRA (South African Bureau of Racial Affairs) newsletter, October 1954, p. 7.
47 F. Barnard, *Thirteen Years with Dr. H. F. Verwoerd* (Johannesburg, 1967), p. 34.
48 W. M. M. Eiselen, 'The supply of and demand for Bantu labour', *Bantu*, 5 (1958).
49 I. Evans, *Bureaucracy and Race* (Berkeley, 1997), pp. 65–6.
50 *Ibid.*, p. 66.
51 Address by A. S. Marais in IANA, 'Record of proceedings of first annual conference, 1952', p. 7.
52 Rheinallt Jones, commenting on Marais' Presidential Address, in *ibid.*, p. 10.
53 J. L. Sadie, 'The political arithmetic of the South African population', *Journal of Racial Affairs*, 4 (1950), 3.
54 *HAD*, vol. 66, 1949, cols 1667–8.
55 See e.g. Union of South Africa, *Senate Debates*, vol. 2, 1950, col. 2206.
56 W. Eiselen, 'The demand'; Union of South Africa, *Report of the Department of Native Affairs for the Year 1951/2*, UG 37/1955, p. 20.
57 Verwoerd in *HAD*, vol. 77, 1952, cols 774–5.
58 Johannesburg Non-European Affairs Files, A 10/3 vol. 1, W. J. P. Carr, Manager of Non-European Affairs to Mr. H. Goldberg, re 'Methods adopted by the Non-European Affairs Department, Johannesburg City Council, in assessing Bantu population figures'.

A MANIA FOR MEASUREMENT

59 M. C. Botha, 'Present and future trends in Bantu Administration', IANA, 'Proceedings of the 11th Annual Congress, 1962', p. 88.
60 Johannesburg Non-European Affairs Files, A 78/2, Department of Bantu Administration and Development, 5/7/68: General Circular no. 16 of 1968: Physical Planning and Utilisation of Resources Act.
61 In respect of the tenants: 'Name, Identity Number, House Number and Township, Number of Minor Children residing with her, Number of Adult Children residing with her, Number of Dependents Residing with her, Whether or not Employed, Whether widowed, divorced, single, deserted', Influx Qualifications, National Unit'; in the case of the third and fourth categories, additional questions were added concerning the economic activity engaged, type of business, address of business. (A 78/2, Johannesburg Non-European Affairs Department, Head Office, to All Senior Township Superintendents, re 'Settling of Non-Productive Bantu Residents in European Areas, in the Homelands', 1/12/69.)
62 This list is compiled from Republic of South Africa, *Report of the Department of Bantu Administration and Development for the Period 1 January 1971 to 31 December 1971*, RP 41/73.
63 Republic of South Africa, *Department of Bantu Education, Annual Report for the Calendar Year 1965*, RP 55/1966, p. 1.
64 HAD, vol. 18, 1966, col. 3569.
65 Hacking, 'How do we do the history of statistics', p. 194.
66 Confidence in the authority of statistics and the measurability of power was not confined to the state alone, which had a significant effect on the manner of political debate and criticism within the country. Statistical discourse was, more widely, a discourse of power. In much of the parliamentary debate about the merits and demerits of apartheid, opposition MPs cited the country's demography as 'proof' that apartheid was not working, while National Party supporters produced different figures showing the rate at which apartheid had succeeded in slowing the rate of African urbanisation. Also, the South African Institute of Race Relations, one of the most vocal liberal critics of apartheid, chose to present its most sustained social and political commentary in the form of an annual 'survey of race relations'. The idea that race relations were best analysed and understood by measuring them indicated the extent to which the state's epistemological assumptions about statistical measurement were more widely shared within the polity.
67 Union of South Africa, Bureau of Census and Statistics, *Annual Report of the Statistical Council, 1951, Incorporating the Report of the Director of the Bureau of Statistics*, p. 6.
68 Republic of South Africa, Bureau of Statistics, *Annual Report of the Statistical Council and the Secretary for Statistics for the year 1 January to 31 December 1969*, p. xi.
69 For example, a comparison of teacher-pupil ratios across the different types of 'Bantu' schools can be extracted from the heaps of data contained in the Department of Education's tables; but the Department itself omitted to do so.
70 For example, the Department of Bantu Administration and Development instructed local authorities to compute counts of 'Unproductive Bantu' using a series of categories which overlapped – viz. 'Registered female tenants', 'registered female lodgers', 'registered professionals', 'registered industrialists and traders'. Occupational status and residential status were not mutually exclusive, making the contents of the table thoroughly mysterious and unreliable.
71 Department of Bantu Administration and Development, Arbeidsvoorligtingsbrief no. 18 van 1969, 25 June 1969, re 'Registrasie van Reisende Bantoewerkers'.
72 The realities of urban labour markets, and interventions by labour bureaux, are discussed in detail in Posel, *The Making of Apartheid*, chs 3, 5, 6, 7.
73 E.g. Dr D. Ziervogel, 'Die buitelandse Naturel in Suid Afrika', in *Die Naturellevraagstuk – Referate gelewer of die Eerste Jaarvergadering van SABRA* (Stellenbosch, 1950); Sadie, 'Some notes on Bantu demography'.

74 Department of Native Affairs, 'Verslag van die Interdepartmentele Kommittee insake Ledige and Nie-Wekended Bantoes in Stedelike Gebiede' (Botha Report), 1962, para. 13.
75 Johannesburg Non-European Affairs Files, A 10/3 vol. 1, W. J. P. Carr, Manager, Non-European Affairs Department, Johannesburg, to H. Goldberg, re 'Methods adopted by the Non-European Affairs Department, Johannesburg City Council in assessing Bantu population figures', 10 June 1963.
76 *HAD*, vol. 87, 1955, col. 524.
77 Johannesburg Non-European Affairs Department, A 78/7/3, 'Toestroming van Bantoes na die geproklameerde gebied van Johannesburg', n.d.
78 In recognition of this problem, urban administrators adopted as their rule of thumb that 'illegals' constituted about 20 to 25 per cent of the urban African population. But this was more properly an estimate as there was no way of reliably quantifying the degree of error in their capacities for population measurement.
79 J. M. Coetzee, 'The mind of apartheid: Geoffrey Cronjé', *Social Dynamics*, 17:1 (1991), 27.
80 Dubow, *Scientific Racism in Modern South Africa*, p. 248.
81 Thomas, *Colonialism's Culture*, pp. 15–16.

CHAPTER SIX

Police dogs and state rationality in early twentieth-century South Africa[1]
Keith Shear

One afternoon in the 1940s, a sub-headman in the Transkei's Tabankulu district was killed with a spear near the Umzimvubu River, some distance from his home. Neighbours found his body the following morning. The police came and, making no progress, instructed the local headman to summon everyone in the area to assemble the next day at the murder scene, where in the meantime the body was to remain unburied. At this gathering, now two days after the stabbing, the police, returning with two dogs, ordered the people to sit in a large circle around the corpse, from which the handler gave the dogs scent before releasing them. The dogs circled round for some time, sniffing each person in turn, until both eventually jumped up on and barked at a man who quickly confessed and was led away by detectives.[2]

In my informants' memory, the dogs' employment in these events stood out as emblematic of Africans' experiences of the white-supremacist state. Yet their account also points to the types of intra-communal tensions, exacerbated by the broader systemic pressures of colonialism, which led to police dogs being brought in and through which the dogs' actions were popularly interpreted. The murder victim had reallocated a desirable field belonging to his assailant. In an era of stock limitation, fencing and forced relocation resulting from the state's 'betterment' proposals for the crowded African 'reserves', land was an increasingly coveted resource and the cause of much resentment, particularly against those responsible for its apportionment.[3] Such resentment cost the sub-headman his life.

The police professed little interest in these allegedly parochial concerns. Having elicited a confession, and justifying their unorthodox methods with disparaging references to Africans' 'mentality', they were content to consider the 'crime' 'solved'. Such complacency, while disclosing an easy predisposition to coercion, also implied the existence of social and moral complexities that the police feared entangling themselves in. Police racism masked nervousness about the limitations of official knowledge and control of social dynamics in African communities as much as it registered arrogant certainty.[4] The police may have known that the introduction of dogs would result in a confession, but they understood only poorly the social phenomena that produced this outcome.

By the time of the Tabankulu events, similar incidents involving police dogs had been occurring in the South African countryside for decades. The police represented dogs as a modern scientific investigative technology which could help to negotiate dealings between rulers and ruled on terms of the authorities' choosing. But the expectation of the dogs' truth-telling effects upon black suspects was largely a matter of faith. The dogs simplified and integrated the semantics of colonial interaction, permitting officials to circumvent court-imposed evidentiary burdens and the uncertainties of employing black police, informers and interpreters, whose reliability they were suspicious of. Yet officials' inability fully to comprehend and control all the circumstances of the 'smelling-out' rituals they staged also enabled Africans to construe these events in quite unexpected ways and to appropriate them for altogether unintended ends.

The purpose of this chapter is not merely to document one more example of how white South African authorities abused their black subjects, but rather to interrogate claims about the modernity and effective reach of the South African state. The theme of police dogs is appropriate to such an enquiry, for work with these animals in African communities exemplified the disjunction between officials' insistence that the South African Police (SAP) apotheosised rational scientific policing and the tawdrier reality of investigations that, as in the case of the murdered sub-headman, more closely approximated witch-finding ordeals. That canine trailing and identification possessed this intrusive character certainly reflected, at one level, the brute coercive capabilities of the state. But that such investigations took this particular ritualistic form also suggests that they were shaped by the authorities' incomprehension of the moral politics of the African communities which they policed and, by implication, the limits of their power.

Historically, from the Old South to Nazi Germany, dogs have been a favoured instrument of repressive regimes. Colonial and metropolitan police forces cooperated with one another in developing this coercive apparatus. In the interwar decades, the SAP took pride in forging such partnerships and in alerting its counterparts internationally to novel uses of dogs. Police dogs have been employed in liberal political orders too. In the United States in the 1990s, civil rights organisations investigated and sued Canine (commonly called K-9) Units following reports of people being needlessly bitten during discriminatory police deployments of dogs in minority neighbourhoods.[5]

In South Africa, although the authorities have employed trained dogs for decades, critical studies of policing neglect the issue, possibly because it has seemed a mere diversionary detail among the totality of iniquitous practices and technologies of racial domination.[6] From their owner's perspective, however, these dogs were far from insignificant. At a ceremony in January 1972 at Cato Manor, Durban, the site of a once vibrant African community which suffered forced removal under apartheid, the Commissioner of Police, with the SAP Chaplain leading the assembled in prayer, unveiled a memorial, 'Helderus' ('Heroes' Rest'), in honour of fallen police tracker and patrol dogs.[7] The memorial was a late instance of a much longer police contribution to the making of dogs as powerful symbols of settler control in colonial Africa – a symbolism registered in an extensive southern African folklore about 'racist' dogs and creatively explored further afield in the fiction of the Kenyan writer Ngugi wa Thiong'o.[8]

First experimented with in South Africa by an amateur enthusiast in the Natal Police in 1909, police dogs were again tested in 1911 in the Transvaal and then gradually introduced throughout the Union. Until the 1960s, dogs were not used in urban patrolling but were employed exclusively for tracking, predominantly in rural districts.[9] Yet, unlike in England and the United States, tracker dogs were not used chiefly to hunt known fugitives. The idea was certainly mooted. In 1906, during a scare following several widely publicised attacks on white farmhouses by deserting Chinese indentured miners, the mining houses proposed the creation of a canine unit to help recapture their workers. 'Blood-hounds are used in other parts of the world for tracking criminals, and there seems to be no reason why a kennel or kennels should not be established on the East or West Rand, whence the dogs could be obtained for service at short notice.'[10] But the Transvaal government apparently did not pursue the suggestion.

Rather, from the outset, the police in South Africa used dogs mostly to discover unidentified perpetrators in unsolved cases – investigative means that produced highly questionable judicial results. In the early years the Crown obtained numerous lower-court convictions 'on the sole evidence of the dogs having pointed out the culprit'. The consequences for the accused could be severe. In 1918, a guilty finding on a charge of stealing a sheep, based 'almost entirely' on 'the behaviour of the police dogs and [on] their tracking of the accused', led a Transvaal magistrate to sentence two men each to eight lashes and twelve months' imprisonment with hard labour. Such evidence was also admitted in jury trials.[11]

A few sceptical senior police officers in these early years protested that it was 'a very dangerous procedure' to make arrests exclusively on dog-trailing information, but headquarters in Pretoria overruled them.[12] Mostly, police and prosecutors claimed a 'scientific' status for canine evidence.[13] *Rex (R.) v. Kotcho*, an eastern Cape case of stock theft, is illustrative. Testifying before the magistrate, the dogmaster, Nel, recalled arriving at the scene with two dogs from Grahamstown. One dog, given scent from some barefoot marks, proceeded to Kotcho's home. Entering, it sat on some blankets and barked.

> I took the accused, another native, and Willie Zendlova with me. I placed the three of them 30 yards from the trail. They had to sit down 30 yards apart. I again put the bitch on the scent of the bare-foot spoors. The bitch then went off to the three natives. She smelled first the other two boys. When she got to the accused she put her paws on him and commenced barking, indicating that he was the man who made the spoor. I then took her off and closed her eyes. I changed the three boys about. I then put her on the scent of the bare-foot spoors, and after that she again picked the accused out and behaved similarly as in the first instance . . . I then took the dog. I gave him the scent of the bare-foot spoors. He also picked out the accused and behaved exactly as the bitch did.

In this testimony, distances are precisely recorded, an impression of geometric symmetry is created, deliberate precautions seemingly exclude a predetermined outcome and the results are successfully replicated. Nel's account exuded the language of controlled objective experimentation to a degree frighteningly revealing of the intrusive, even torturous, nature of the process. The Solicitor-General later argued that Nel, having 'studied as a science the movements and habits of police dogs', rendered 'expert' testimony, the dog being 'merely an instrument'. In other cases, handlers supported their technical expositions with plans and diagrams.[14]

In 1918–19, however, the canine programme experienced a serious legal challenge when several cases (including *Kotcho*'s) reaching the Supreme Court resulted in rulings that police dog identifications were inadmissible as evidence in trial. Chief Justice Innes, in a 1919 Appellate Division decision in the case of *R. v. Trupedo*, wrote disparagingly of 'the super-canine sagacity claimed for these animals' by 'their optimistic instructors' – claims that lent 'to such evidence a dangerously exaggerated importance' prejudicial to the accused. Innes thus echoed the concerns of the sceptical minority within the police force itself. He denied that one could infer with 'scientific or accurate knowledge' from a police dog's barking or jumping up at someone that this was the same person whose scent it had been instructed to trail. Such inferences involved 'conjecture and uncertainty'. Canine evidence was 'analogous to hearsay' – inadmissible because the dog (the immediate witness) could not be cross-examined. This was so irrespective of whether the dog in question was of a breed reputed to have acute powers of smell, was 'of pure blood', possessed a particularly discriminating nose, was highly trained in tracking, and had been placed on a trail indisputably the perpetrator's – considerations that had led some American states' courts, in decisions quoted as precedent in the South African cases, to hear testimony on these points as a 'foundation' for admitting dog-trailing evidence in trial.[15] The onus of proof, the Chief Justice implied in R. v. *Trupedo*, properly belonged to Rex the Crown, not Rex the dog.

Yet the courts' censure of police dogmasters' claims to scientific expertise did little to hinder the expansion of the canine programme in the interwar decades. For, despite ruling dog-tracking evidence inadmissible, the judges endorsed the animals' use as an investigative technique for producing other evidence that could be led in a trial. As Sir Thomas Graham, Judge President of the Eastern Districts Local Division, put it, inadmissibility did 'not in any way seriously interfere with the employment by the police of these animals. They may still ... be usefully employed for the purpose of obtaining clues ... Once a clue is discovered relevant evidence is usually forthcoming which can be duly produced.'[16] Graham probably did not foresee that the 'relevant evidence' increasingly would take the form of dog-induced confessions. Ironically, given Innes's apparent insistence that police and prosecutors' evidence should meet strict standards of relevance, the courts' endorsement gave legal recognition to the programme's end results while removing from legal scrutiny any trace of canine involvement in the production of these results. Henceforth the Crown would need to observe the courts' requirements only in a formal sense.

In countenancing a separation of the process of procuring from that of proving evidence deriving from the use of dogs, the judiciary freed the canine programme's inner workings from effective external supervision. Meanwhile, the silencing of internal sceptics heeding a more scrupulous legalism removed a further significant check on the SAP's readiness after 1918 to privilege investigative procedures that were less methodical and rational than those that a previously greater police commitment to satisfying court-imposed evidentiary standards in a substantive and not merely formal way had exacted. The Crown's earlier conscientiousness in its approach to court-centred rule was a consequence less of benevolence towards the ruled than of the capacity of the legal process to foster more methodical bureaucracies and the rationalisation of society in the era of state-building and economic expansion following the South African War. But by the end of the First World War, popular resistance, especially black political militancy, was compelling administrators to calculate the costs in commitment of personnel and resources that continued adherence to a more substantive legalism as a basis for enhancing bureaucratic instrumental rationality would exact. Officials' reservations encouraged their retreat into an institutional logic favouring the entrenchment of far less rational mechanisms of rule such as the canine programme exemplified. Rationalising forces increasingly were channelled into differentiated spheres of state activity sealed off from judicial scrutiny, producing a more brutal but less methodically assertive interwar state.[17] Officials now argued that their ability to maintain effective state control demanded governmental standards which were at odds with liberal metropolitan ideals. Difficulties, however, arose when whites' liberties as well as blacks' were curtailed. The solution – sometimes with the courts' connivance – was to differentiate still further the supposedly colour-blind common-law legal domain from the administrative sphere.

That racial calculation informed the courts' bifurcated approach to the canine programme was revealed in Graham's decision in *R. v. Kotcho, R. v. Barley*. 'In the[se] two cases', the Judge President wrote, 'the charges are of stock theft, and the accused are natives, but if it is held that evidence of this nature is relevant and thus admissible, the rule will apply to cases of every description, and respectable citizens of every class in life may have their lives, liberty and reputation jeopardised by evidence that a trained police dog had indicated them to be guilty persons'.[18] In dividing the processes of discovering and proving evidence derived from dogs, the courts handed the police the discretion to employ investigative means in dealing with blacks that they rarely used in connection with whites. In practice this discretion mainly affected the staging of scent identification parades, for inevitably there were occasions when trails from crime scenes unexpectedly led the dogs to white suspects.[19]

Yet, details that the Commissioner of Police, I. P. de Villiers, revealed in the late 1920s about the dogs' training confirmed that a racist orientation also informed tracking. The dogmasters used an African to lay a 'night trail' in the countryside surrounding the training camp the day before a practice session. The man went into hiding and the next day the dogs were exposed to an article bearing his scent and instructed to find him. '[I]t is not long', the account crowed, 'before he is discovered by his pursuers, who pounce upon him, barking to attract attention'. Such recitations of canine accomplishment, while aiming self-consciously to be scientific, were unwittingly thoroughly anthropomorphised and suffused with white-supremacist prejudice. 'The dogs readily respond and take a keen interest in their work', this same source asserted.

Racist settler ideas about Africans' hygiene and bodily odour were put to work too. The training staff, like most southern African whites, assumed that Africans collectively possessed a particularly intense and enduring odour. Thus police officials readily claimed their dogs could 'pick up the scent of tracks seventy-two hours old'. De Villiers did not think it tested the credulity of even sympathetic readers to cite a case in which two dogs allegedly successfully trailed a stock thief for fifty miles from six-day-old footmarks. Such essentialist themes coexisted effortlessly with the programme's fundamental assumption that all humans, Africans included, left individual scent trails that dogs could distinguish. 'Different natives are employed [in training]', the account continued, 'so that the animals do not become accustomed to any particular quarry.'[20] To round out their preparation, the dogs were taken to 'native locations where practice trails are worked to accustom them to such conditions' – as though the dogs, left to their own devices, might contemplate such spaces with the same horrified bewilderment that characterised whites' descriptions of Africans' living arrangements.[21]

Clearly, although the late 1910s' court decisions may have pricked the police's pretensions to 'scientific' knowledge of canine behaviour, they left considerable space for the development of expertise about the breeding, training, care and handling of police dogs. Indeed, the sealing off of the investigative domain (in which admissible evidence was procured) from the scrutiny of the legal arena (in which such evidence was proved) shielded the technique from searching adverse criticism and permitted the police to continue to maintain that they were following scientifically controlled procedure. Statistics were diligently recorded for each case of the distance the dogs trailed and of the age of the scent they were pursuing.[22] The Commissioner's interwar annual reports are replete with details of canine investigations recounted in the same language of objective experimentation that the dogmaster Nel employed in *Kotcho*. Rationalising zeal, instead of being directed into purposive investigation, was put to the service of thoroughly irrational state practices.

Thus the courts, far from seriously obstructing the use of dogs for judicial purposes, ironically assisted a major expansion of the canine programme. When the lease on the site of the dog-training depot at Irene expired in 1922, new facilities were erected at Quaggapoort, six miles west of Pretoria.[23] The growth in the number of police dogs and of cases in which they were tried was particularly marked during the 1920s, rising from 65 dogs based at ten stations and used in 541 cases in 1920, to 202 dogs at thirty stations and used in 2,044 cases in 1929. In only about fifteen per cent of cases were handlers able to give the dogs scent from the scene that led them to a suspect – a record in the light of which officials' increasing resort to the animals reflects a faith in canine potential that is all the more remarkable.[24] Dogs became an integral and familiar element of rural policing in interwar South Africa.

The interwar Police Commissioners, T. G. Truter and I. P. de Villiers, both assiduously fostered the programme. Truter victimised subordinates who questioned the dogs' efficacy while promoting others who favoured their employment.[25] In his reports Truter saluted 'the wonderful trailing powers of the dogs', their 'almost phenomenal' success in detecting stock thieves, and 'the great moral effect that the sagacity of the dogs has on the native mind'.[26] Only a dearth of suitable handlers, in Truter's opinion, limited the programme's extension. 'So insistent is the call for these animals from many places in the Union', he enthused in the mid-1920s, 'that it has been difficult to find sufficient men with the requisite temperament and keenness to undertake the care, training, and handling of these useful adjuncts to criminal investigation.' Trainees at Quaggapoort were encouraged to believe 'that the work performed by dogs was absolutely genuine' and to perceive themselves as part of an exclusive fraternity.[27] Truter's passion did not escape the wit of Minister of Justice Tielman Roos. When his senior permanent secretary suggested the department need no longer see regular reports because all scepticism had been vanquished, Roos agreed but enjoined him not to 'hurt Compol's feelings by making him think that we disparage his dogs'.[28]

De Villiers was as ecstatic about the programme as his predecessor. In April 1929, shortly after becoming Commissioner, he contributed material to the London periodical *The Police Journal* for an article, 'Dogs as detectives in South Africa'. A formal claim to international leadership in the field, the article announced the arrival of dogs as 'a distinct section of the Police Force' in South Africa and emphasised that Quaggapoort was 'the only state-owned institution of the kind in the world'. The article described the curriculum for handlers and dogs at the depot, the care and breeding of the animals, and several criminal cases in which different breeds and cross-breeds were 'successes'.[29]

Time only reinforced de Villiers's early enchantment. 'These dogs', he wrote in 1934, 'undoubtedly are a very potent factor in reducing stock thefts.' 'The prevention and detection of stock theft is almost entirely dependent upon the work of the trained Police dog', he claimed on one occasion. 'No police force operating in rural areas,' he later insisted, 'can be regarded as complete without a police dog section.' De Villiers argued hard for resources to back these assertions: a trained dog was worth £80 and each station was 'an expensive business', requiring at least two dogs, an expert handler and a motor car and driver 'for the exclusive conveyance of dogs and the trainer'. By 1934 de Villiers had increased the number of stations countrywide to thirty-six, some serving areas as large as 30,000 square miles.[30]

As the canine programme matured, it garnered wider acclaim. Within the police, with the sceptics silenced, the junior ranks shared their Commissioners' almost mystical faith in the dogs – an enthusiasm imparted to the white South African public as a symbol of its own nationhood and a vindication of its racial supremacy in magazine articles admiring the animals 'as a scientific arm of the Police'.[31] Although officials opposed scent-discrimination demonstrations lest publicised failure should damage the prized 'moral influence' the dogs 'produce[d] on the native mind', canine acrobatics later joined police physical training and horsemanship displays as popular highlights at agricultural shows, trade exhibitions and other occasions for patriotic advertisement.[32] Members of Parliament with large farming constituencies repeatedly demanded more extensive use of police dogs in the 1920s, whereas in the immediate post-Union period the topic had occasioned mainly back-bench mirth in the House of Assembly.[33] This shift in sentiment reflected the growth of white South African nationalism, the corresponding coarsening of interwar white political discourse and the diminishing expectation of routine bureaucratic efficiency. The 1926 South African Police Commission, describing the training and use of police dogs as a 'specialised form of criminal investigation work', extolled the programme as 'a most important arm' of the Criminal Investigation Department (CID). White farmers, the Commission reported, were 'unanimous' in praising not only the dogs' success in detecting stock thieves, 'but particularly' the magical 'deterrent effect which their employment ha[d] on the mind of the Native stock thief'. Farmers, believing Africans to be 'very much afraid' of police dogs, readily attributed declines in crime to the dogs' presence in a district.[34]

Canine work thus epitomised the SAP's self-image of rational scientific superiority and mastery; it became a distinct specialisation in which the force claimed regional and international leadership, and with which it calculatedly disseminated some of its more obnoxious technologies and doctrines. 'The present police dog kennels and training camp outside Pretoria', a retired officer exalted in the mid-1930s, 'are probably the most modern and up-to-date to be found anywhere.'[35] British officials in neighbouring Basutoland, Swaziland and Bechuanaland allowed Union police dogs to pursue cross-border scent trails and called on the SAP for canine assistance in their own cases.[36]

Between 1918 and 1939 the SAP received enquiries from the authorities in Southern Rhodesia, Kenya, Palestine, Burma, Australia and India about training handlers and supplying dogs. De Villiers pointedly refused 'to despatch trained police dogs to another Force in the Empire unless [he] had had an opportunity of fully training at least two dogmasters of that Force'.[37] Southern Rhodesia sent a Corporal Ansell for training in 1927, but he failed to complete the course. Policemen from Palestine arrived next in 1934 and soon after deployed their dogs in the 1936 Arab general strike.[38] Much chauvinistic fanfare attended such exchanges. 'Quaggapoort', bragged a 1938 SAP promotional booklet aimed at white schools, 'has acquired such a reputation for its thoroughness of Police Dog training and the efficiency of its dogs, born, bred and trained there, that policemen from other parts of the British Commonwealth of Nations are sent here for training as Dogmasters; and, when trained, they take back to their own countries South African Police dogs for police work there.'[39] In the metropole, too, some cognisance was taken. In 1935 a former Superintendent of the Yorkshire West Riding Constabulary commended the programme as 'an example to the Mother Country which one would like to see followed now that authentic incidents proving the dog's value are frequently happening'. Four years later the *San Francisco Times* printed an appreciative newswire under the headline 'Afric Police Dogs Lead the World'.[40]

The identification of canine work with criminal investigation already defined it as a sphere of exclusive expertise in the opinion of SAP management. 'All our dog-training', de Villiers boasted in a 1938 reply to an enquiry by K. D. Wagstaffe, of the CID in Peshawar, India, 'is designed to use the dog as a detective and in this respect we differ from every other country where the dog is used as a protection for the policeman.' In seeming desire to protect the SAP's monopoly in this arcane science, the final version of de Villiers's letter deliberately omitted a detailed account, previously appended to a draft, of the curriculum followed at Quaggapoort.[41]

The atmosphere of exclusivity in which the SAP enveloped its canine work was redolent of South Africa's racial caste system. A dogmaster, characterised by de Villiers as 'a first-class policeman skilled in the detection of crime, a lover of dogs, and above all patient and steady under adverse circumstances', was necessarily white in the eyes of police management – a conception the SAP insisted upon in sharing canine expertise with overseas police forces.[42] This insistence on white expertise resulted in an awkward correspondence between Peshawar and Pretoria which illustrates how South African officials could make their own brand of racist practices prevail elsewhere.

Answering de Villiers, Wagstaffe regretted that his North-West Frontier Province Police had too few 'European Sergeants' to spare any for training as dogmasters, but declared that he could send two commissioned 'Indian Officers' who 'would of course conform in every way to European customs'. A. F. Perrott, Wagstaffe's chief, wrote personally to reassure Pretoria that he would 'select well educated men of good family and good manners who would fit in anywhere, provided that they were given a fair chance. This I am sure would be the case,' he added dubiously. His doubts were confirmed in Chief Deputy Commissioner George Baston's cool response to Wagstaffe. 'Your desire', said Baston, 'to safeguard the dignity of your Officers is fully understood and appreciated by me, as I am sure my difficulty in the matter is by you.' 'Our Dog Training Depot', Baston expanded separately to Perrott, 'is manned entirely by Europeans, and in this country, where the colour bar is so clearly defined, it would be quite impossible for me to accept your Indian Officers for a course in dog training and mastership.'[43]

Perrott should have known, Baston implied, that 'mastership' of anything in segregationist South Africa was a white preserve. De Villiers clearly had promoted his expertise effectively, however, for Perrott soon buried his scruples, pronouncing himself 'so fully convinced of the value of Police dogs' that he was 'prepared to make almost any sacrifice to get two men trained'. Shortly afterwards two white non-commissioned officers from India were warmly received at Quaggapoort. They were offered 'any amount of relaxation, sports, etc.', and left six months later with 'pleasant memories' of South Africa, having become 'efficient Dogmasters... capable of training young dogs' and taking with them '6 dogs... carefully selected... from [the SAP's] best strain'.[44]

The specialised nature of canine work was marked not only by the elaborate eighteen-month curriculum developed to teach dogs to distinguish and track individual scents, by the rarefied qualities sought in potential dogmasters, and by their own lengthy apprenticeship in the handling, care and training of dogs. As the reference to the 'best strain' indicates, breeding police dogs was a precious science too. Police headquarters initially bought all its dogs from professional breeders in Europe. In the interwar years, although continuing occasionally to import 'thoroughbreds... from Europe for the purpose of counteracting inbreeding', the police took pride in meeting its needs locally. During the 1920s headquarters carefully studied 'the peculiarities of the various types of dogs' to determine which were the 'most suitable ... for the South African country and climate'. Indeed, the SAP's official history celebrates the emergence of 'the true South African police dog' from domestic experimentation.[45]

The officer most responsible for raising breeding to a 'science' was a future Commissioner of Police, R. J. ('Bobby') Palmer, who claimed that as commandant of Quaggapoort in the early 1930s he had transformed a previously 'haphazard and indiscriminate' system of breeding. In ironic recognition of police dogs' growing reputation as 'detectives', Palmer, as a pen-profile later put it, 'from training dogs ... was promoted to training men' as commandant of the Police Training Depot in Pretoria in the mid-1930s.[46] Palmer's replacement at Quaggapoort had few reservations about his predecessor's achievement. All the SAP's dogs were bred at the Depot itself with 'flattering results', he enthused in 1938. '[W]e select outstanding trained dogs from a line of ancestors famed for brains, grit and vitality, endowed with a strong scenting nose'. The staff experimented with Rottweilers, Bloodhounds, Airedales, Alsatians, Rhodesian Ridgebacks, Pointers and various cross-breeds, but settled on the 'pure and well-bred Dobermann-Pinscher' as the 'most useful and reliable' police dog. Alsatians, however, were 'quite unreliable and ... not recommended'.[47] In this way, the discourse of breeding harmonised with officials' predilection for racial categorisation and ranking of people.

The language of science was mobilised to describe other elements of the project. The section on the programme in the Commissioner's annual reports exhibited discourses on veterinary medicine and canine occupational illness.[48] Elsewhere the Commissioner publicised the dogs' treatment by 'veterinary experts' and mentioned that their diet included a 'special cake manufactured' at Quaggapoort.[49] Veterinary and dietary concerns also infiltrated speculation about the effects of climatic and soil conditions on scent trails. 'Scent', remarked an extraordinary document produced at Quaggapoort in 1928,

> is an effluvium which is constantly issuing from the pores of all animal substances ... it may be said to depend chiefly on two things, the condition of the ground and the temperature of the air ... It lies badly with a North or East wind ... Fog, as a rule, is bad for it, as is also frost ... It lies best in the richest soil ... Failure in following up scent is sometimes due to the dog's olfactory organs being affected. This will frequently be found to arise from colds, constipation, or other causes, which a dose or two of opening physic seldom fails to remove. A little Sulpher [sic] or Syrup of Buckthorn will generally have the desired effect.[50]

Shielded from the scrutiny of the courts and from internal police dissent, every facet of the interwar canine programme could be similarly invested with the trappings of scientific respectability and precision. These 'sciences', however, were not merely discursive veneers of modernity, but also manifestations of instrumental rationality canalised into what can only be considered 'irrational' practices when judged by the standards of modern scientific superiority the police themselves claimed to exemplify.

Amid all the self-congratulation, a major commission of inquiry into the police in the late 1930s voiced a more cautious appreciation. Its report dutifully genuflected to the dogs' value in deterring stock theft, lauded the 'excellent arrangements and the skilful methods of training' at Quaggapoort and cited 'a very favourable report by the Inspector General of Police in Palestine' on dogs obtained from the SAP – thus sustaining the force's claim to international leadership. But the Commissioners, luminaries of the legal establishment, also hinted darkly at potential inquisitorial bias in the dogs' employment. Although quickly denying any 'dishonesty', their report mentioned 'evidence... of a disturbing nature... tending to show that a dogmaster can, if so minded, control the activities of a dog by secret signals in such a way as to make the dog point out any particular person suspected by the dogmaster'. Coupled with the belief that Africans thus picked out were likely to confess purely out of fear, such 'secret' skills could undermine equitable criminal procedure.[51]

Yet, from the investigators' perspective, the value of staging scent-identification parades lay, as Sir Thomas Graham had observed in 1918, in the hope that these events would yield 'clues' leading to legally relevant admissible evidence. In practice, foremost among such 'clues' were confessions produced by terror in the face of sniffing dogs. Indeed, an experienced handler, who came to believe police dogs were 'a hopeless failure', implied that Africans whom dogs picked out at parades very often confessed, even when other evidence suggested their innocence.[52] In a murder case a confession could entail the death sentence.[53]

Police reports on the 'successful' employment of dogs unintentionally confirm how arbitrary and disquieting these incidents appeared to Africans. In a 1934 case of attempted safe-breaking in the south-eastern Transvaal, scent from a piece of piping at the scene led the dog to a nearby compound where it barked in the corner of a room sleeping twenty workers. The dogmaster paraded all 150 compound residents in three rows. The dog pointed out a man in the third row who promptly confessed but also asked why the dog had selected him rather than his 'accomplice'.[54]

Police dogs frequently trailed directly into domestic spaces, entering a room or hut at night and sniffing each of its sleeping occupants in turn, which eight Africans experienced in a 1919 investigation of a break-in in the western Transvaal town of Zeerust; rushing right up to and pointing out one person among several at a homestead, as happened in a 1925 case of stock theft in the Cape's Middelburg district, and in a 1936 murder case in Natal's Ladysmith district; or jumping on a compound bed and barking, as in a 1941 stock-theft investigation in the Transvaal's Ermelo district.[55] The less corroborating evidence and the more inconclusive the scent trail, the larger appears to have been the size of the 'picking-out parade'. When a saddle disappeared from an airfield stable in 1945 and the dog lost the scent on a paved runway leading towards some black soldiers' barracks, the police turned out 600 Africans in twelve rows for a parade in which the dog pointed out 'a native corporal'.[56]

The courts, as we have seen, afforded no protection against these intrusive and demeaning proceedings. If a dog followed a scent to a hut or room, this was often deemed sufficient cause to enter and search at once without a warrant. In 1926 the Cape Provincial Division of the Supreme Court set an important precedent in dismissing an action of damages for *injuria* against the government brought by a Kokstad 'Griqua' cheese factory worker; police dogs had incorrectly pointed him out as the thief of a missing cheese by putting their paws on his shoulders and barking during a parade of the factory's employees. In the original summons the worker contended that the dogs' acts were 'an aggression ... upon [his] liberty and freedom and an indignity and insult'. The sense of affront is unmistakable even through such stiff legalese. In an earlier suit relating to this incident the worker complained that detectives had threatened to set the dogs on him unless he confessed and that he had been bitten and scratched upon refusing.[57] The dismissal of his action meant that henceforth the police could organise their terrifying dog identification 'line-ups' with impunity.

Black South Africans undoubtedly experienced police dogs' 'smelling out' of alleged 'criminals' as oppressive. But we should resist a simple victimisation analysis of how the confessions were elicited. Certainly police exhibited brutal violence, but their increasing reliance on the dogs in the interwar period also suggests a more limited state whose functionaries were withdrawing from commitments to more rational and methodical forms of administration. Officials made no serious effort to understand the mechanism's operation. 'There is a mystery,' the Kenya Police Commissioner reported following a 1927 visit to Quaggapoort with his Union counterpart Truter, 'something perhaps uncanny or undefined to the native mind which the use of dogs produces, and in a number of cases the use of dogs has brought the wrong doer forward to confess to his crimes. The Commissioner assured me that the moral influence obtained by the use of dogs was very [great] as far as native criminals are concerned.'[58] Clearly, more was happening than can be exhaustively explained either by fear of police violence or by officials' complacent stereotypes about Africans' mentality. Africans doubtless did fear the dogs, but the content of that fear is not self-evident. Thus, although this chapter's main point is to analyse officials' notions of what they thought they were achieving, their infatuation with magical effects, their limited comprehension of cause and consequence, and what this implies about the interwar South African state's modernity and reach, it is important to attempt a plausible, albeit conjectural, account of how social processes within African communities contributed to producing the confessions.[59]

In form, the scent identification parades closely resembled witch- and witchcraft-finding practices. It is possible that Africans found in these state-sanctioned rituals ready substitutes for these outlawed practices. For in the Union, as elsewhere in colonial Africa, the proscription of imputations and detection of witchcraft left Africans feeling vulnerable to evil from within their own communities in ways that new colonial pressures and opportunities probably exacerbated but that secular courts could not address.[60] The line-ups, after all, commonly comprised people known to each other from a variety of daily interactions. The social unit could be a farm or other workplace, a compound, location, a large household, or the followers of a particular headman. The parades were thus subject to local understandings and observances, tested loyalties, and became occasions at which enmities could be vented. Africans to whom police dogs initially trailed not infrequently remained silent until the dogs had singled them out a second time from among their neighbours in a parade, which suggests a social ritual dimension to their confessions. Nor was it unusual for those indicated then to implicate others, which intimates that these events were forums for expressing parochial conflicts and jealousies only tangentially related to the cases the police were investigating.[61]

What local social understandings would Africans have brought to these events? Ethnographies indicate that dogs occupied an ambiguous position in the beliefs of many African communities in early twentieth-century South Africa.[62] As domesticated animals, dogs were valued for hunting, herding and keeping watch at night. Unlike cattle or goats, they were seldom sacrificed. The anthropologist Monica Hunter found in Pondoland that people believed dogs drove away *uthikoloshe*, the best known of witches' familiars. Yet their proximity to humans also meant that dogs were potential sources of danger. B. A. Marwick observed in Swaziland that a dog entering a hut during childbirth was not removed until the baby had ridden on its back to deter future misfortune. Dogs were suspected familiars in their own right, bearing 'a message of malice' from their owners, especially if seen jumping on to the roof of another person's hut, or found urinating inside another's home or cattle enclosure. In isiXhosa, as in other languages, the verb 'to smell', *ukunuka*, also represented the action of diviners in revealing the sources of witchcraft. Dogs that actively sniffed at people were particularly distrusted as likely familiars. Most intriguingly suggestive is A. T. Bryant's recounting of a tradition in which Shaka's 'magic dog' settled a chieftaincy dispute by picking one of the claimants out of a parade.[63]

These ideas may well have informed Africans' perceptions of police dogs. From this perspective, the dogs were not only producing perpetrators for the 'crimes' their handlers happened to be investigating, but simultaneously revealing sources of conflict or evil within the community. Alternatively, the dogs were thought to be bewitched, bringing misfortune upon those they picked out or whose domestic spaces they violated. In either case there was a yawning gap between what the authorities thought to effect and the unintended consequences of their interventions in the context of the complex dynamics and cleavages of parochial politics.

Truter, infatuated with 'the great moral effect' he believed his dogs produced on Africans, observed uncritically that 'in a great many instances native offenders have been so surprised at the uncanny knowledge possessed by the dogs in following them up when they have considered themselves absolutely untraceable and free from arrest, that they have confessed to their crimes.'[64] White officials happily accepted these dog-induced confessions as an opportunity to close their files, but Africans' confessions were as likely indicative of communal moral tensions sharpened by the material exigencies of colonialism. The case cited at the outset of the sub-headman murdered in an era of increasing landlessness exemplifies such tensions and exigencies. As anthropologists have noted, confessions by those accused of witchcraft, to which the confessions at dog smelling-out parades bear a striking resemblance, have historically been surprisingly common; people who confessed when pointed out were acknowledging feelings of resentment that they genuinely believed had caused others' misfortune.[65] Thus the police were able to meet the formal condition for admissibility of confessions into court proceedings that they be 'freely and voluntarily' given, although the whole parading ritual was essentially an ordeal.[66]

From the police's perspective, the availability of dogs as an ostensibly 'scientific' technology allowed investigators to objectivise crime within a preferred epistemological framework positing a direct relationship between a narrowly defined 'criminal' action and an alleged perpetrator. Obviating complicating questions requiring methodical investigation, the use of dogs reduced communicative interaction between rulers and ruled to a level with which the police authorities felt comfortable. What officials could not recognise is that the very inertness they so valued in the medium afforded Africans interpretive possibilities of their own. Thus we have the ironic spectacle of a state institution celebrating its modernity and international leadership through a technique that Africans locally may well have been appropriating for ritual ends condemned as 'irrational superstition' and prohibited as 'uncivilised' by this same state.

The discourse of modernity legitimating the canine programme rested ultimately on two irrationalities governing state institutional action itself.[67] First was the faith in the dogs' powers, which the 'scientific' languages of breeding, training, veterinary medicine, olfaction, etc., never entirely disguised. Second, in the domain of policing practice, the conditions for means-end rationality were only superficially approximated in the serendipitous correspondence of investigative aims and closed-case results. What occurred in between was far less perfectly understood and drew officials on to terrain where the procedural legitimations of modern states were inoperative. The short cut to satisfactory outcomes that the canine ritual facilitated thus represented an effective qualification of bureaucratic instrumentality's colonisation of early twentieth-century South African society. The substantially greater pre-1918 official punctiliousness about procedural niceties that initially had significantly promoted bureaucratic ubiquity and the rationalisation of society had given way by 1939 to a less ambitious state whose personnel rarely scrupled about their extensive reliance on a mechanism of rule employing dogs and magic.

Notes

1 Research for this chapter was assisted by a grant from the Joint Committee on African Studies of the Social Science Research Council and the American Council of Learned Societies with funds provided by the Ford, Mellon and Rockefeller Foundations. I am grateful to many colleagues who engaged with earlier versions at conferences and seminars in Durban, Brighton, Chicago and Ann Arbor. For their

detailed comments, my particular thanks to David William Cohen, Saul Dubow, Jonathon Glassman, Randall Packard and Lynn Thomas. Unless otherwise indicated, manuscript sources are from the Central Archives Depot in Pretoria.
2   S. Mthakasi and E. Vava interviewed by J. Mzayifani and K. Shear, 4 August 1994.
3   W. Beinart and C. Bundy, 'State intervention and rural resistance: the Transkei, 1900-1965', in M. A. Klein (ed.), *Peasants in Africa: Historical and contemporary perspectives* (Beverly Hills, 1980), pp. 298-304.
4   For a skilful decoding of African resistance registered in settlers' racism, see K. E. Atkins, *The Moon is Dead! Give Us Our Money! The cultural origins of an African work ethic, Natal, South Africa, 1843-1900* (Portsmouth, 1993).
5   J. R. Lilly and M. B. Puckett, 'Social control and dogs: a sociohistorical analysis', *Crime & Delinquency*, 43 (1997), 123-47; Amnesty International, *United States of America: Torture, ill-treatment and excessive force by police in Los Angeles, California* (London, 1992), pp. 29-34.
6   Critical studies include G. Cawthra, *Policing South Africa: The South African police and the transition from apartheid* (London, 1993); and J. D. Brewer, *Black and Blue: Policing in South Africa* (Oxford, 1994).
7   See 'Helderus onthul', *Sarp*, 8:5 (1972), 28-9; and C. R. Stanley, *Deferred Value: An hitherto unpublished account of the origin and development of the British police dog* (Chichester, 1978), pp. 28-9, also cited in Lilly and Puckett, 'Social control', p. 138. These authors fail to note the location's significance.
8   Ngugi wa Thiong'o, *A Grain of Wheat* (London, 1967).
9   M. de W. Dippenaar, *The History of the South African Police, 1913-1988* (Silverton, 1988), pp. 41, 297; M. Hansen, 'Dogs on the beat', *Justitia*, 1:10 (1962), 3; A. W. Brink, 'Die totstandkoming van die S. A. Polisiehondeafdeling', *Sarp*, 1:10-11 (1965).
10 Barlow Rand Archives, Sandton, H. Eckstein & Co., 253, File 134 ('Labour'), Part 2, No. 771, 'Chinese labour. Suggestions for securing further efficient control of Chinese labour, for the apprehension of deserters, and prevention of crime', 28 March 1906.
11 Archives of the Police Inquiry Commission, 1936-37 (K80), vol. 21, 15 December 1936, evidence of Captain Donald, pp. 1551-3; *Rex (R.) v. Adonis*, 1918 Transvaal Provincial Division (TPD) 411 at 411-12; *R. v. Kotcho, R. v. Barley*, 1918 Eastern Districts Local Division (EDL) 91 at 96-8; *Star*, 13 December 1917.
12 Archives of the South African Police (SAP), 1/3/23/2: Deputy Commissioner, Johannesburg, to Secretary, SAP, Pretoria, 31 December 1917 and 12 January 1918; Secretary, SAP, to Deputy Commissioner, Johannesburg, 7 January 1918; K80, vol. 21, 15 December 1936, evidence of Captain Donald, pp. 1543-8.
13 Although the technique's veracity is immaterial to this study, it may be noted that the 'scientific' evidence for dogs' capacity to distinguish and trail individual human scents was disputed throughout the twentieth century and remains inconclusive. See W. Craig, 'The dog as a detective', *Scientific Monthly*, 18:1 (1924), 38-47; and I. L. Brisbin and S. N. Austad, 'Testing the individual odour theory of canine olfaction', *Animal Behaviour*, 42:1 (1991), 63-9.
14 *R. v. Kotcho, R. v. Barley*, 1918 EDL 91 at 93-4 and 97 (citing the 1918 Transvaal case of *R. v. Moheketse*, mentioning plans of scent trails).
15 *R. v. Trupedo*, 1920 Appellate Division (AD) 58 at 61-3; *R. v. Adonis*, 1918 TPD 411 at 413-14. The fullest local exposition of the American authorities was Eastern Districts Judge President Sir Thomas Graham's in *R. v. Kotcho, R. v. Barley*, 1918 EDL 91 at 98-102, to which the judges in *Adonis* and *Trupedo* deferred. For discussion of the still evolving American case law, see B. Finberg, 'Annotation: evidence of trailing by dogs in criminal cases', *American Law Reports*, 3d, 18 (1968), 1221-40 and Supplements. See also S. G. Chapman, *Police Dogs in North America* (Springfield, 1990), pp. 67-9. English precedent had greater standing with South African judges: the absence of relevant English authorities signified to both Graham and Innes that it had not been 'considered proper or right to lead evidence in [English courts] of law as to [dogs'] actions'. See *Trupedo* at 61; *Kotcho, Barley* at 95-6.

POLICE DOGS AND STATE RATIONALITY

16 *R.* v. *Kotcho, R.* v. *Barley,* 1918 EDL 91 at 105. See also *R.* v. *Adonis,* 1918 TPD 411 at 413.
17 For elaboration and substantiation of this argument, see K. S. Shear, 'Constituting a state in South Africa: The dialectics of policing, 1900–1939', Ph.D. dissertation, Northwestern University, 1998.
18 *R.* v. *Kotcho, R.* v. *Barley,* 1918 EDL 91 at 104.
19 *Annual Report of the Commissioner of Police* [hereafter *ARCP*] *for 1935* (An. 606-'36), pp. 13–14; SAP 36/29/42, Officer Commanding (OC) SAP Dog Depot to Commissioner of the South African Police (Compol), 20 January 1942, para. 8.
20 'Dogs as detectives in South Africa, prepared from material supplied by Colonel I. P. de Villiers, M.C., Commissioner, South African Police', *Police Journal,* 2 (1929), 189, 191–2. On settler ideas about Africans' hygiene and bodily odour, see T. Burke, *Lifebuoy Men, Lux Women: Commodification, consumption, and cleanliness in modern Zimbabwe* (Durham, 1996), pp. 20–1.
21 SAP 21/20/38, OC SAP Dog Depot to Compol, 16 July 1938.
22 *Ibid.,* Annexure 'A'. On officials' cult of statistics see D. Posel's chapter in this volume.
23 *ARCP for 1922* (UG 9-'24), p. 54.
24 *ARCP,* 1920–29. A third to a half of the dogs were assigned to stations; the remainder were kept in training or in reserve at Quaggapoort. 'Success' statistics were given for 1920 to 1927 and each year the Commissioner felt obliged to justify the small percentage. Significantly, later reports omitted this information.
25 K80, vol. 21, 15 December 1936, evidence of Captain Donald, p. 1553.
26 *ARCP for 1923* (UG 15-'25), pp. 53–4; *ARCP for 1926* (UG 7-'28), p. 15.
27 *ARCP for 1925* (UG 6-'27), p. 68; K80, vol. 71, 2 April 1937, evidence of Constable Barnard, p. 5871.
28 Archives of the Secretary for Justice, 1/140/25, handwritten minutes by Bok and Roos, 9 January 1925, on Acting Commissioner, SAP, to Secretary for Justice, 8 January 1925.
29 'Dogs as detectives', 188–92.
30 SAP 1/190/31, de Villiers to Secretary for Justice, 8 February 1934; *ARCP for 1928–29* (UG 13-'30), p. 12; SAP 21/20/38, de Villiers to K. D. Wagstaffe, Assistant to the Inspector General of Police, CID, North-West Frontier Province, India, 25 July 1938 (draft); 'Dogs as detectives', 188; SAP 21/6/49, OC SAP Dog Depot to Compol, 5 May 1939. Frequently, the distance to the crime scene meant that two or more days elapsed before the dogs began trailing, but this did not diminish investigators' confidence in the evidence produced.
31 'How four-footed detectives work: a morning with the police dogs', *Picture News* (January 1932), 22–3, magazine enclosed in SAP 21/14/49.
32 For opposition to public trailing demonstrations, see SAP 21/199/26, Commissioner, Kenya Police, to Colonial Secretary, Nairobi, 'Police Dogs – South African Police', 10 September 1927, copy enclosed in Office of the Commissioner, Nairobi, to Colonel Truter, 13 September 1927. On mounted police and physical training displays, see *Star,* 12 April 1930.
33 *House of Assembly Debates* (*HAD*): 1912, col. 119; 1924 (1st Session, 5th Parliament), col. 720; 1929 (1st Session, 6th Parliament), col. 588.
34 *Report of the Commission of Enquiry to Enquire into the Organisation of the South African Police Force Established under Act No. 14 of 1912* (UG 23-'26), p. 13; K80, vol. 69, 30 March 1937, evidence of William Ernest Sinclair Moor, p. 5552.
35 R. S. Godley, *Khaki and Blue: Thirty-five years' service in South Africa* (London, 1935), p. 242.
36 *ARCP for 1931–32* (An. 79-'33), pp. 14–15; *ARCP for 1932* (An. 84-'33, 2nd Session), pp. 21–2; *ARCP for 1938* (An. 515-'39), p. 10.
37 SAP 21/20/38, de Villiers to K. D. Wagstaffe, Assistant to the Inspector General of Police, CID, North-West Frontier Province, India, 25 July 1938 (draft).
38 *Star,* 11 December 1937; Stanley, *Deferred Value,* p. 11.

39 Archives of the Public Service Inquiry Commission (K47), vol. 13, H1/2A, enclosing Union of South Africa, *The South African Police as a Career* (Pretoria, 1938), pp. 5-7.
40 Ex-Superintendent R. Arundel, 'Police dogs', *Police Review*, 28 June 1935, 627; *San Francisco Times*, 22 August 1939, cutting in SAP 21/6/49.
41 SAP 21/20/38, de Villiers to Wagstaffe, 25 July 1938 (draft copy, with certain paragraphs marked 'omit' in margin).
42 *Ibid.*
43 SAP 21/20/38: Wagstaffe to de Villiers, 20 August 1938; A. F. Perrott, Inspector-General of Police, North-West Frontier Province, Nathiagali, India, to de Villiers, 24 August 1938; G. R. C. Baston, Chief Deputy Commissioner, SAP, to Wagstaffe, 5 October 1938; Baston to Perrott, 5 October 1938.
44 SAP 21/20/38: Perrott to Baston, 7 January 1939; Baston to Perrott, 30 January 1939 and 16 September 1939; Baston to Inspector-General of Police, North-West Frontier Province, Peshawar, India, 14 October 1939.
45 *HAD*, 1912, col. 119; *ARCP for 1934* (An. 673-'35), p. 12; 'Dogs as detectives', 189-90; Dippenaar, *History*, p. 41.
46 'General Palmer's Life Story – II', *Cape Times Magazine*, 11 August 1951; 'Personality Parade', *Indaba*, August 1954. Cuttings in album kindly shown to me by Mrs C. McLennan, daughter of R. J. Palmer.
47 SAP 21/20/38: OC SAP Dog Depot to Compol, 16 July 1938; de Villiers to Wagstaffe, 25 July 1938 (draft).
48 E.g. *ARCP for 1924* (UG 21-'26), pp. 51, 67.
49 'Dogs as detectives', 189.
50 SAP 1/3/23/2, Commandant, SAP, Quaggapoort, to Compol, 16 October 1928, enclosing memorandum 'Work of police dogs: scent'.
51 *Interim and Final Reports of the Commission of Inquiry to Inquire into Certain Matters Concerning the South African Police and the South African Railways and Harbours Police* (UG 50-'37), pp. 68-9; K80, vol. 71, 2 April 1937, evidence of Constable Barnard, p. 5876.
52 K80, vol. 71, 2 April 1937, evidence of Constable Barnard, pp. 5871-9.
53 *ARCP for 1930* (UG 35-'31), p. 9. With serious crimes like murder a confession rarely sufficed to convict. See Criminal Procedure and Evidence Act, No. 31 of 1917, sec. 286. In the lower courts, which dealt with stock theft and other offences with which Africans were charged as a result of canine investigations, convictions on mere confessions were readily obtained: since most defendants were unrepresented, the admission of their confessions went unchallenged. Although these courts heard less serious criminal cases, they could nonetheless impose severe sentences.
54 SAP 21/20/38, OC SAP Dog Depot to Compol, 16 July 1938, Annexure 'A', para. 1. Despite the confession, the magistrate refused to convict in this case: *ARCP for 1934* (An. 673-'35), p. 13.
55 R. v. Trupedo, 1920 AD 58 at 60; *ARCP for 1925* (UG 6-'27), p. 70; SAP 21/20/38, OC SAP Dog Depot to Compol, 16 July 1938, Annexure 'A', para. 4; SAP 36/29/42, OC SAP Dog Depot to Compol, 20 January 1942, para. 12.
56 *ARCP for 1945* (UG 27-'46), p. 5.
57 *Mentor v. Union Government*, 1927 Cape Provincial Division (CPD) 11 at 13, 15; *Union Government v. Mentor*, 1926 CPD 324 at 325; *Star*, 9 July 1926. The incident gained international notoriety when the government claimed it was not responsible for wrongs committed by policemen performing their statutory duty.
58 SAP 21/199/26, Commissioner, Kenya Police, to Colonial Secretary, Nairobi, 'Police Dogs – South African Police', 10 September 1927, copy enclosed in Office of the Commissioner, Nairobi, to Colonel Truter, 13 September 1927.
59 On the uses of 'conjectural history' see T. C. McCaskie, 'Accumulation, wealth and belief in Asante history', *Africa*, 53 (1983), 25-6.
60 M. Hunter, *Reaction to Conquest: Effects of contact with Europeans on the Pondo of South Africa* (London, 1936), p. 275; M. Chanock, *Law, Custom and Social Order: The colonial experience in Malawi and Zambia* (Cambridge, 1985), ch. 5.

POLICE DOGS AND STATE RATIONALITY

61 SAP 21/20/38, OC SAP Dog Depot to Compol, 16 July 1938, Annexure 'A', para. 9; SAP 36/29/42, OC SAP Dog Depot to Compol, 20 January 1942, para. 4; SAP 21/6/49, Compol to Secretary for External Affairs, 5 April 1944, enclosing 'Work of police dogs: extract from annual report, 1943', para. 3; *ARCP for 1945* (UG 27-'46), p. 6.
62 On animals' ambiguous cultural status generally see E. Leach, 'Anthropological aspects of language: Animal categories and verbal abuse', in E. H. Lenneberg (ed.), *New Directions in the Study of Language* (Cambridge, Mass., 1964), pp. 23–63. For dogs' ambiguous this- and other-worldly aspects in Kongo thought and their part in witch-finding, see W. MacGaffey, 'The eyes of understanding: Kongo Minkisi', in *Astonishment and Power: Kongo Minkisi and the art of Renée Stout* (Washington, 1993), pp. 39–43. My thanks to J. Glassman and L. Thomas for these references.
63 A. I. Berglund, *Zulu Thought-Patterns and Symbolism* (Cape Town, 1976), pp. 284–5; A. T. Bryant, *Olden Times in Zululand and Natal Containing Earlier Political History of the Eastern-Nguni Clans* (London, 1929), p. 482; A. T. Bryant, *The Zulu People* (Pietermaritzburg, 1949), p. 327; H. Callaway, *The Religious System of the AmaZulu* (Springdale, 1870), p. 28, n. 53; S. S. Dornan, 'Dog sacrifice among the Bantu', *South African Journal of Science*, 30 (October 1933), 628–32; Hunter, *Reaction*, pp. 287, 297; E. J. Krige, *The Social System of the Zulus* (Pietermaritzburg, 1936), pp. 189, 325; A. Kropf and R. Godfrey, *A Kafir-English Dictionary* (Lovedale, 1915), p. 295; B. A. Marwick, *The Swazi: An Ethnographic Account of the Natives of the Swaziland Protectorate* (Cambridge, 1940), p. 143.
64 *ARCP for 1923* (UG 15-'25), p. 54.
65 D. Hammond-Tooke, *Rituals and Medicines: Indigenous healing in South Africa* (Johannesburg, 1989), pp. 82–3.
66 Criminal Procedure and Evidence Act, No. 31 of 1917, sec. 273.
67 M. Weber, *Selections in Translation* (Cambridge, 1978), pp. 28–9. Measuring the procedures of officials who claimed to exemplify 'rationality' against their own professed standards does not mean these were objective standards whose employment by officials to assert Africans' 'irrationality' had any validity.

CHAPTER SEVEN

The Race Welfare Society: eugenics and birth control in Johannesburg, 1930–40[1]

Susanne Klausen

Only the nobler, more intelligent, energetic and healthier citizens of the present should be the ancestors of future generations.[2] (H. B. Fantham, from a lecture presented to the Race Welfare Society, 14 August 1930)

The State needs all the *good* children it can get. But it does not need the under-nourished, mal-adjusted brood of the overworked and under-paid slum dweller or the feebleminded, and so the woman eugenist would extend to them the knowledge of birth control.[3] (Annie Porter, from a lecture given at the same meeting; emphasis in original)

Introduction

On 26 June 1930 thirteen women and men gathered together in Johannesburg to form the Race Welfare Society, the purpose of which was 'the study, investigation and application of eugenics with especial reference to South African problems'.[4] The founders of the Race Welfare Society (RWS) turned to eugenics for a blueprint on how to cultivate a healthy and productive white population. If implemented, they believed, eugenic measures would eliminate hereditary disease and poverty among whites, and mitigate white maternal and infant morbidity and mortality.[5] Most important, eugenics would solve the perceived problem of differential birth rates within the white population which resulted in a faster rate of reproduction among 'poor whites' than among the middle class, a phenomenon said to be causing white 'race degeneration' and 'feeblemindedness'.[6] Preventing the propagation of 'inferior' whites through eugenics, according to the RWS, would improve the quality of the 'race' as a whole; ensure the maintenance of white supremacy in South Africa well into the future; and ease the tax burden on the middle class who resented subsidising state social welfare initiatives for 'poor whites'.

Eugenics was the science and social movement dedicated to improving physical, mental or moral qualities in human populations. Widely prevalent in industrialising countries during the first third of the twentieth century, it was an expression of the biological worldview of modern society.[7] Eugenics movements took root in many different parts of the world and under varying political systems. They can therefore only be fully analysed and understood with reference to the particular historical contexts in which they operated.[8] Nevertheless, a common feature of eugenics movements everywhere – including in Johannesburg – was a naive faith in the socially transformative power of science. Indeed, the legitimacy conferred upon the RWS by its leaders (who counted prominent scientists among them) was of central importance to its success. Research on eugenics in South Africa has begun to emerge in recent years which elucidates connections between scientific racism and the entrenchment of white supremacy, as well as the gendered nature of eugenic discourse.[9] But there has been little written about the relationship between birth control and eugenics and, by comparison with European and North American historiographical traditions, very little concerning the politics of reproduction more generally.[10]

The RWS's first public meeting held six weeks after it was formed was a considerable success: seventy people attended, a constitution was adopted and an executive committee selected, consisting mostly of highly educated male professionals.[11] Anxious to intervene in a practical way in social problems exacerbated by the Great Depression, the RWS's executive committee set out to apply eugenic theory, and birth control was immediately seized upon as the most promising opportunity to do so. As the RWS's 1934 annual report stated, 'Members hardly need to be reminded that the Race Welfare Society is eugenic in intention and that its main purpose is to secure the physical and mental betterment of the race. To achieve this aim the most practical course appears to lie in the wise direction of the Birth Control movement.'[12]

Between 1932 and 1937, the RWS opened five birth control clinics in downtown and working-class neighbourhoods in Johannesburg, three for white and two for black women.[13] Moreover, it initiated an impressive propaganda campaign promoting modern, medicalised birth control at public lectures and symposia which were frequently reported in prominent newspapers. These events resulted in significant popular support for both eugenics and birth control in the early 1930s. The RWS also secured assistance from social welfare agencies in locating poor women deemed appropriate for contraception and successfully lobbied various levels of government for financial support. By the end of the decade, the RWS had done much to transform birth control from a shameful moral issue to a 'public health' issue and a 'maternal health' service in Johannesburg. Furthermore, the RWS was pivotal in the formation of a national birth control coalition in 1935 which promoted birth control on a national level.[14]

This paper examines the RWS and analyses its ideology, leadership, goals and accomplishments, as well as its relationship to various levels of the state.[15] The RWS experienced two distinct phases in the 1930s. From 1930 to 1933 it was a reactionary organisation characterised by an overriding commitment to preventing the reproduction of 'inferior' whites. After 1934, liberals replaced key conservatives who had departed, and as a result the RWS's highly conservative eugenic focus softened somewhat to include concern for maternal health among women of *all* 'races'. Though this chapter is organised in two parts in order to reflect this periodisation, the line dividing them is not intended to be drawn too sharply since there was considerable overlap and consistency in membership and ideology throughout the decade.

The early history of the RWS, 1930–33

Exemplifying the pervasiveness of hereditarian ideas, eugenists in the first third of the twentieth century tried to put biological theories about 'racial' health to practical use, either through selective breeding or environmental modification, or else a combination of both. 'Positive' eugenics encompassed those measures targeted towards increasing desirable traits in a population whereas 'negative' eugenics sought to eliminate so-called 'dysgenic' (biologically harmful) traits. During the 1930s the RWS's agenda was dominated by negative eugenics as was evident in attempts to limit the fertility of 'poor whites', the 'feebleminded', those with inherited disease and the unproductive members of white society through measures including segregation, sterilisation and birth control. Conversely, positive eugenics existed to a much lesser extent, mainly in exhortations to 'better quality' (middle-class) whites to produce larger families. In addition, a philanthropic impulse was represented within the RWS by the desire to mitigate poverty and suffering, and to improve women's health, through curbing the fertility of the poor. From the very beginning, the RWS attracted adherents who often subscribed to more than one of these objectives; this resulted in an alliance of interests among its leadership on the executive committee.

From 1930 until 1933, under the guidance of H. B. Fantham (1876–1937) and to a lesser extent his wife Annie Porter (1880–1963), the RWS was a highly conservative organisation. Both were respected scientists who enjoyed considerable professional reputations and who readily drew upon the authority of 'objective' science to buttress their social prescriptions. Fantham was a professor of zoology and comparative anatomy at the University of Witwatersrand and he was the founder of the department of zoology and dean of the faculty of science from 1923 to 1926.[16] He had an international reputation as a eugenist and represented South Africa at the International Federation of Eugenic Organisations in 1931.[17] He was also chair of the RWS and president of the Eugenics Society of South Africa, based in Pretoria, of which he was also a founding member.[18] On his own and in collaboration with Porter, Fantham produced numerous scientific papers on the heritability of physical and 'racial' characteristics. They presented their findings to the South African Association for the Advancement of Science and elsewhere throughout the 1920s and early 1930s.[19] Porter established the department of parasitology at the South African Institute for Medical Research in 1917 and was its head until 1933. She was also a senior lecturer in parasitology at the University of the Witwatersrand.[20] She was dedicated to projects which affected public health to the extent that she once deliberately infected herself with tapeworm in the course of a study on measly meat at abattoirs.[21]

Fantham and Porter were outspoken in their opposition to the interruption of 'natural selection' by putatively naive philanthropic and social welfare measures, arguing that such interventions acted to preserve the 'unfit' in society. Doing so, they insisted, merely aided social undesirables to reproduce their kind. In Fantham's words:

> In the earlier stages of civilised society, man interfered little with nature's method of selection, the result being that the incapable, foolish, mentally or physically weak died out or were eliminated by the tribe. To-day, philanthropic measures, preventive measures against disease, better medical services have resulted in the rearing to maturity of many individuals, inadequate mentally, physically and socially, and allowing them to reproduce offspring as inadequate as themselves.[22]

Fantham was an extreme biological determinist who believed that 'Mental and moral differences are almost entirely due to the influence of heredity and ... are but slightly affected by environment'. Porter's biological determinism was just as steadfast: 'Philanthropists', she said, 'fall all over themselves to help the afflicted and obviously unfit. We have cripple schools, deaf and dumb schools, special schools for the mentally disturbed, for the truant, the thief and for all sorts of unfit. In these costly special schools, every effort is made to do the often impossible and turn the inadequate into adequate citizens'.[23] Both subscribed to eugenics in the belief that it would ensure 'natural selection' was replaced by the 'rational selection' of superior qualities in a population. As highly educated professionals they felt certain that they, and their like-minded followers, possessed the requisite skills and knowledge to determine who qualified as worthy of selection and who did not. Fantham in particular devoted a large proportion of his eugenic rhetoric to outlining the dangers of 'poor whiteism', 'feeble-mindedness' and 'hybridism' (miscegenation) which, he believed, produced 'tainted stock' – hereditarily defective persons who should not be allowed to propagate. 'Eugenists', he declared, 'desire that the unfit shall not procreate and that their taint shall die with them'.[24]

Despite Porter's, and particular Fantham's, early conversion to eugenics and steadfast attempts to promote it, they garnered little popular support during the 1920s. However, the Great Depression (1929–33) and attendant social and economic crises created a more amenable ideological climate. During these years South Africa experienced great economic and social disruption and hardship: total income from industrial manufacturing dropped by 20 per cent, export prices plummeted, bankruptcies were common, and 22 per cent of white and Coloured males were officially unemployed (unemployment statistics for women are unavailable).[25] In Johannesburg, white male unemployment rose over 400 per cent from 17,000 in 1926 to 72,000 in 1932–33, and among Africans endemic poverty was intensified by the state's so-called 'civilised labour policy' of replacing black labour with unemployed whites.[26] Starving in 'Native Reserves', large numbers of Africans flooded to the cities in the early 1930s in search of livelihoods in the growing manufacturing industry with the result that, by 1936, they outnumbered whites.[27] Faced with massive numbers of poverty-stricken unemployed workers, rapid African urban in-migration, and heightened labour militancy, middle- and upper-class South Africans perceived that their society – and their privileged position within it – was under threat. Among those who eschewed a social analysis of the upheaval around them, some responded by developing a biological interpretation of the situation (the proliferation of the 'unfit') and advocating a 'scientific' response (eugenics).

Key members of the RWS's executive committee included social conservatives as well as respected social welfare reformers. A representative of the former was Edward Mansfield whose interest in birth control was motivated solely by a commitment to negative eugenics. It was he who instigated the formation of the RWS and recruited Fantham and Porter.[28] Mansfield was closely involved in establishing the RWS's first birth control clinic in 1932 and was central to the formation of the South African national birth control coalition in 1935. In contrast, Henry Britten (1878–?) and J. L. Hardy (1879–1941) had long been active social reformers prior to joining the executive committee. Britten, chair of the RWS from late 1932 until 1944, was Chief Magistrate of Johannesburg until he retired in 1932 and thereafter pursued full-time his self-described hobby of social welfare work, a passion that appears to have been bound up with his long-standing involvement in the Anglican Church. Britten was awarded a Coronation Medal in 1937 for his philanthropic work.[29] Through his ongoing involvement with the Rand Aid Society, he was well aware of the extent of white unemployment and poverty in Johannesburg: In 1930 alone the Rand Aid Society received over 200 requests each month for employment from newly arrived whites to the city.[30] Hardy was the RWS's treasurer and auditor.[31] He was also a founder member of the Johannesburg Board of Charities and was active in the South African Institute for Race Relations as well as other social agencies interested in 'Native' welfare.[32] For example, he took a leading role in the establishment of the Rand hostels for boys where juvenile delinquents could find housing and employment. His long-standing involvement in efforts to combat juvenile delinquency probably stimulated his interest in birth control: In view of the perception that 'delinquents' were created by the problem of unstable homes, he believed that limiting the size of poor families would lead to better parenting and fewer troubled children.

Doctors were well-represented in the RWS leadership from the beginning. At the 1932 annual general meeting, eight of the fifteen members elected to the executive committee were doctors.[33] From the beginning, non-medical members of the executive committee firmly subscribed to the medicalisation of birth control and never questioned the assumption that contraceptive services should only be made available to 'patients' (as clients were termed) in clinics under medical supervision. Consequently, the executive committee actively sought out doctors, and the approval of the South African Medical Council before it engaged medical staff to devise clinic procedures, manage clinics and train District Surgeons in contraceptive techniques.[34] (The RWS sent pamphlets to all District Surgeons in Transvaal and the Orange Free State inviting them to visit its birth control clinic and requesting that they send eligible – namely poor, white and married – women to Johannesburg for contraception.[35]) Sympathetic doctors could also help win over the support of the medical profession and the state by helping to legitimise contraception.[36] This was especially valuable given the highly controversial nature of birth control in South Africa during the early 1930s. While the Anglican Church had ended its opposition to birth control (when practised within marriage) at the 1930 Lambeth Conference, the Catholic and Dutch Reformed Churches remained adamantly against it on moral grounds. (The Dutch Reformed Church did not reverse its position until 1934.[37]) Moreover, various powerful Afrikaner nationalists such as D. F. Malan, who was the Minister of Public Health from 1924 to 1934, opposed the establishment of birth control clinics for political reasons. Seeking to increase the *volk* rather than limit it, male nationalists were against providing access to contraception; indeed, they strongly endorsed the gender ideal of the *volksmoeder* whose social role was to reproduce the *volk* in both biological and cultural terms. Another likely reason for Malan's resistance was the perception that the RWS's disgust for 'poor whites' – who were embraced by Afrikaner nationalists – could not be countenanced. However, nationalists' apprehension of birth control diverged along gender lines. M. E. Rothmann, the Organising Secretary for the Afrikaans Christian Women's Union in the Cape, the sole female investigator on the Carnegie Commission of Inquiry Into the Poor White Problem (1929–32), and a staunch Afrikaner nationalist, publicly endorsed birth control in 1933 as an aid to rural *volksmoeders*. She was persuaded of women's need to control their fertility as a result of her vast personal contact with rural 'poor white' women in the late 1920s and early 1930s, and her sympathy for such women who were forced to bear children under extremely difficult and dangerous conditions is palpable in her volume for the Carnegie Commission.[38] Finally, in the early 1930s there was still widespread fear among many middle-class whites that birth control would contribute to 'depopulation' (white 'race' extinction), a belief that held much currency in the wake of the national 'black peril' election of 1929 and rapid urbanisation of Africans.[39]

For their part, doctors joined the executive committee for a variety of reasons not least of which was to bring birth control under the control of the medical profession. In 1934 Dr I. Block from the RWS explained to the South African Medical Council at its annual meeting that birth control was more than a private matter for the people involved, it was also a public health issue and as such 'demands the participation and guidance of the medical profession'.[40] He and other doctors were also convinced that widespread access to contraception would reduce the frequency of illegal abortion with its associated dangers, and improve the economic position of the poor.[41] Others such as Dr A. Marius Moll, a member of the executive committee 1930–32, appear to have been solely interested in negative eugenics and motivated by the desire to eliminate 'defective stocks'.[42] The high proportion of female doctors who worked in a volunteer capacity in RWS birth control clinics suggests that they may have felt drawn to the birth control movement out of a sense of feminist solidarity but no direct evidence to support this supposition has been uncovered. Certainly, birth control afforded female doctors with family responsibilities an opportunity to continue practising in their profession outside the high demands of a general practice, since birth control clinics held only one or two sessions per week.[43] Furthermore, work at the clinics, which by 1937 were staffed exclusively by women, appears to have been a branch of medical activity that women could largely define, manage and claim as a female domain free from male competition and interference.

The diversity of key members' social and political perspectives was more a question of degree than kind, for four overlapping threads bound them together into a relatively cohesive whole. First, they shared an assumption of white and middle-class superiority which was broad enough, if not uniformly deep, to accommodate both the extreme socially conservative views of someone like Fantham and the liberal wing represented by the likes of Hardy. Second, they held in common a worldview that explained social problems in biological terms, though there was a distinction between those who did so in an intellectually systematic, rigorous fashion and those who were merely inclined to do so as 'fellow travellers'. Third, they wished to mitigate their society's social problems – as they defined them – and finally, as intellectuals and professionals, they shared a faith in the efficacy of science over politics or religion as the arena in which to pursue their goals. A common concern to maintain the existing class and racial order in South Africa was combined in the case of some members with a genuine, deeply felt desire to eliminate human suffering. Whatever their primary motivation, they all believed their aims could be furthered through providing the most needy (and potentially threatening) South Africans with contraception.

Initially, the RWS's faith in eugenics led to an ambitious programme that included objectives encompassing many fields of activity, such as educating its members about eugenics; influencing public policy and legislation in favour of eugenic measures; developing and distributing propaganda about eugenics 'calculated to arouse public interest in and support of Eugenic Education and Reform'; and encouraging and publishing original research.[44] As a result, during its first three years of existence the RWS blossomed from a club for like-minded professionals into an organisation with a significant popular following, demonstrating how deeply its message tapped into white middle-class anxiety: 318 people attended its first annual general meeting in 1931, and about 220 the following year.[45] Over this period public lectures were frequently held on subjects ranging from the ethics of birth control to the inheritance of mental defect, and these events were well-attended and widely reported in newspapers.[46] In addition to race and class interests, gender likely accounted to a significant extent for the early popularity of the RWS. Porter's first public lecture on behalf of the RWS in 1930, 'Eugenics from a woman's point of view', appealed directly to women to join in the eugenics movement by promoting the social role of mothers to a crucially important 'scientific' endeavour. This must have appealed to middle-class, educated and newly enfranchised white women who sought to escape the invisibility and isolation of labouring in the home and to join public life.[47] Indeed, the RWS's focus on 'race improvement' through the provision of medical attention to mothers led far more women than men to join the organisation in subsequent years.

Birth control

During the Great Depression, the problem of 'poor whiteism' grew to worrisome proportions and was a particular source of alarm for eugenists. 'Poor whites' was the widely used term for white Afrikaans-speaking rural *bywoners* (squatters) and petty-commodity producers who were wrenched from rural subsistence by the advent of industrial gold mining after 1886, a series of agricultural crises in the 1890s, the South African War (1899–1902) and the expansion of capitalist agriculture thereafter. Seeking employment in towns and cities, indigent whites streamed into urban areas where they soon constituted a highly visible and politically problematic population. Often living in squalid mixed-race slums and working-class neighbourhoods, 'poor whites' were perceived as a potential threat to the prevailing 'race' and class structure of South Africa. Whether through miscegenation or else through the possibility of an alliance of the poor that transcended racial divisions, 'poor whites' were feared to be undermining the race barrier.[48] The landmark 1932 Report of the Carnegie Commission established the number of 'poor whites' in South Africa at that time to be 300,000 out of a total white population of 1,800,000.[49] By 1932, according to O'Meara, 'the rate of such rural impoverishment and proletarianisation was most rapid in the swiftly changing rural areas of the Transvaal', whose urban centre was Johannesburg where the RWS was based.[50] Poverty among 'poor whites' was often acute and shocking, not least for its tendency to drag 'poor whites' down to Africans' standard of living: one family in Johannesburg was described in 1930 by an Anglican missionary doctor as 'living in native hovels, the father out of work, the mother lying in bed with acute rheumatic fever and heart disease, and six children fed only on mealie meal, one of which lives on a mattress in the corner of the room covered with flies and prostrate with pneumonia'.[51]

Fantham reflected and fostered the fear among middle-class whites that they were in jeopardy of losing their position as 'torchbearers of progress and civilisation' as a result of the rapid proliferation of 'poor whites' and miscegenation between 'poor whites' and blacks.[52] He was extremely distressed to learn that the fall in the white birth rate was located in 'the more educated and better paid classes' while 'poor whites' were 'breeding' uncontrollably, and he transposed his resentment into a discourse on the heredity of social inadequacy:

> In South Africa, large families, often 10 to 12 and even 20 children, occur among the 'poor whites.' It is the slum-dwellers, the feeble in will, the careless, the shiftless and indifferent who have the large families and, consequently, in the future, a larger proportion of the population will have the hereditary characteristics of these classes... The less socially adequate are those largely temperamentally unfit, the careless, inefficient, idle, intractable, slack and self-indulgent. In South Africa we have the 'poor whites,'... forming about eight per cent of the population. These less fit do not sustain their burdens themselves, but they are passed on to the community in unemployment doles, free meals at school, payment of less than their fair share of taxation, etc.[53]

For Fantham, the quality of the white population took precedence over its quantity and, explicit in his desire to eliminate putatively inferior whites for the greater good of the 'race', he declared, 'In South Africa there must be limitation of the "poor white" element'.[54] Other members of the executive committee appealed to middle-class interests by claiming that a lower 'poor white' birth rate would reduce the burden of taxation imposed by government to provide for the diseased and improvident.[55] This kind of self-interest and intolerance towards Afrikaans-speaking indigent whites were the primary motives for establishing a birth control clinic.

Immediately after RWS members unanimously passed a resolution urging the establishment of a birth control clinic 'at the earliest possible moment', the executive committee set out to obtain widespread public support and the approval of municipal and Union governments for its plan.[56] In a letter sent to 600 prospective members in 1931, the executive committee claimed that curbing the reproduction of 'poor whites', the 'feebleminded' ('a scandal and menace to the community') and people with hereditary disease would rid society of undesirable elements, and improve the health and social opportunities of future citizens.[57] In February 1931 the executive committee informed the Department of Public Health of its intention to open a clinic: 'It is hoped', its letter stated, 'by these and allied measures to help ensure healthier bodies and minds for the babies of the future and to diminish the stream of recruits into the ranks of the shiftless and irresponsible section of the population'.[58]

The Secretary for the Department of Public Health, Edward Thornton, was immediately receptive to the executive committee's proposal. By 1931, 'poor whiteism' and white maternal mortality were two pressing political problems facing the central state and Thornton believed they could be mitigated if poor and working-class women limited their fertility.[59] Moreover, he was keen to allow extra-state agencies to distribute contraception so that the Department could evade full financial and, more importantly, political responsibility, a crucial consideration given the strong opposition from various churches and the fear among whites of 'depopulation'. As a result of these obstacles, the Department was not in a position to advocate birth control publicly until the end of the 1930s, though Thornton sought from 1931 onwards to direct volunteer birth control agencies in ways that would suit his purposes. After deprecating the RWS's original proposal he made a counter-offer that would further both his and the RWS's goals, stating that the Department strongly favoured the idea of an 'Outpatient Gynaecological Clinic' where mothers could obtain general medical attention for their gynaecological health needs. One aspect of the proposed clinic's services would be the distribution of contraceptives. Thornton was also persuaded by – or already subscribed to – the RWS's eugenic orientation:

> It seems to the Department that there might be a need in Johannesburg for the development of an Outpatient Gynaecological Clinic where mothers needing advice on physical health matters both of nature and nurture, might be assisted. Such an institution under adequate and skilled supervision could receive minor gynaecological cases, deal with some forms of post-natal aftercare, *accept responsibility for counseling mothers of subnormal physical or mentality*, give advice on contraceptive methods where medically needed and act as a clearing house for appropriate treatment of one sort or another [emphasis added].[60]

Here, for the first time in discussions between the RWS and the Department of Public Health, mention was made of maternal health. Though women had always been the assumed clients for its clinic, the executive committee had not referred to them in its original proposal to Thornton, because, unlike in birth control movements active in the US and England at this time, the RWS lacked a visible and vocal feminist component. At no point did the RWS agitate for women's sexual or social emancipation, and birth control – a crucial tool in women's struggle to obtain greater social freedom – was never apprehended nor endorsed as such; not surprisingly abortion was condemned. Yet the RWS did mobilise public concern over maternal health needs in Johannesburg by drawing sustained attention to the ill health of white mothers who experienced frequent pregnancy and childbirth. Though it was never central to its agenda, the maternal health issue was raised at the earliest RWS events. At the inaugural public meeting, for example, a Dr Bernstein, a member of the executive committee, told the audience it was necessary to seriously consider the 'appalling maternal and infantile death rate' in South Africa which was double that of New Zealand's.[61] The following year, executive committee member Geoffrey Hills, MP for the Labour Party, reported that the working class did not know where to go to obtain birth control advice and as a result 'the health of thousands of women suffered'.[62] Also, Henry Britten announced his belief that 'Women... should have the right to say whether or not birth control should be practised', the only known declaration by a member of the RWS of women's right to control their fertility.[63] Repeated reference to maternal ill-health as an obstacle to white 'race welfare' throughout the decade assisted in highlighting the problem and the need to address it. And in 1936, by which time the RWS was operating clinics for women of all 'races', the organisation became affiliated with the South African National Council of Women, evidence of growing self-recognition that it was an organisation working in the interests of women.[64] In a society lacking an effective women's movement, the RWS's efforts to offer contraceptive services, while rarely justified for women's own sake, were an important source of pressure on decision-makers to improve maternal health. Thornton's 1931 counter-proposal was crucial to broadening the RWS's mandate for birth control to include a central focus on this important issue.

However, Thornton (himself a doctor) effectively ensured that birth control was cast as a medical issue, rather than a woman's issue, by limiting the provision of contraceptives to cases 'where medically needed'. Only doctors would assess which conditions warranted contraception; in addition, relevant 'medical' problems included 'subnormal' physical or mental conditions and again physicians were to be entrusted with the responsibility of diagnosis and contraception distribution. Ultimately, doctors – not women – would have the power to determine who could and could not obtain contraception and in this manner the Department reinforced the medicalisation and 'eugenicisation' processes already well underway within the RWS.[65]

Thornton's letter had a major impact on the executive committee which was most anxious to meet the Department's wishes, and it resolved to amend its strictly eugenic mandate and incorporate all of his suggestions.⁶⁶ For example, the RWS named its first birth control clinic (for whites only) the Women's Welfare Centre, a name that captured Thornton's positive intention to emphasise maternal health rather than the RWS's negative goal of curbing the fertility of the 'unfit'. In fact, throughout the 1930s the RWS's clinics did provide gynaecological health services to all women seeking contraception as a matter of course, including vaginal examinations, referrals to venereal disease and other clinics as necessary, and even advice on overcoming sterility. Yet the executive committee managed to retain its primary emphasis on negative eugenics as is evident in the modified mandate for its first clinic that was resubmitted to Thornton:

> The Centre [will] serve the purpose of a) Counselling mothers of subnormal physique or intelligence, b) Giving advice on contraceptive measures when required in the interest of the Mother's health and the well-being of the family, c) A clearing house for cases of disease requiring medical or surgical care, d) Introducing Mothers to Societies which provide post-natal aftercare.⁶⁷

The decision to accommodate Thornton's suggestions regarding maternal health was one of expediency. The RWS was fixed upon avoiding state opposition to its proposed clinic and hoped, moreover, to obtain a financial subsidy from the Department of Public Health as soon as possible. In essence, broadening the birth control clinic's mandate was instrumental to attaining its overriding goal of delivering contraception to poor and 'feebleminded' whites. The first birth control clinic in South Africa was opened on 4 February 1932 in Sauer's Building on Loveday Street in central Johannesburg.

Consolidation: 1934–44

In 1935 the executive committee started expanding access to medicalised birth control beyond the 'poor white' community to include African, Coloured and Asian women. This was not due to any shift in priority on the part of original members of the RWS's leadership but to the liberalisation of the executive committee as a result of new members who joined in 1934. Three of them were crucial to expanding clinic services to black women: Elsa (Sallie) Woodrow, Winifred Hoernlé and her husband R. F. A. Hoernlé. Woodrow and the Hoernlés immediately utilised their experience in welfare initiatives for blacks by exploring the possibility of extending contraceptive services to Coloured, 'Native' and Asian women. They established a birth control clinic for Indian women in July 1935, and in October 1935 a clinic was opened in the Methodist church on West Street, Ferreira, whose reverend offered the use of his vestry as a birth control clinic for Coloureds only (a clinic for African and Coloured women was opened the following year).

Woodrow (1900-96) was a doctor who had been instrumental in establishing the Cape Town Mothers' Clinic which opened on 15 February 1932, a mere eight days after the RWS's Women's Welfare Centre (neither organisation knew of the other's existence until months afterwards). From the start, the Cape Town clinic was open to white and Coloured women, albeit in segregated sessions, a distinction of which Woodrow long remained proud.[68] She imported her relatively colour-blind approach to birth control to Johannesburg where she resided from 1934 to 1936.

The Hoernlés, both leading academic figures at the University of the Witwatersrand, were co-founders of the South African Institute of Race Relations (hereafter the SAIRR), established in 1929, an important base for advocacy of liberal social and political reforms. Each served terms as its president: R. F. A. Hoernlé in 1934-43 and Winifred Hoernlé in 1948-50 and again in 1953-54. Winifred Hoernlé (1887-1960) soon emerged as a leader of the RWS and became the first woman to chair the organisation in 1944. She was pivotal to the liberalisation of the RWS and provided it with a strong humanitarian impulse. A noted anthropologist, she worked hard to alleviate the hardship facing various social groups in South Africa, indigent children, Africans, Indians and prisoners, and she joined a number of maternalist organisations that addressed black and white child welfare.[69]

R. F. A. Hoernlé (1881-1943) was appointed honorary president of the RWS from 1937 until his death in 1943. He was a professor and chair of the department of philosophy at the University of Witwatersrand from 1924 until 1943 and was made Dean of the faculty of arts in 1925. One recollection notes that 'the native problem was Hoernlé's dominant interest in his Johannesburg years' and as a liberal humanist he wished to 'move the Europeans to understand the natives'.[70] Hoernlé's interest in contraception dates back to at least 1926 when he participated in a controversial symposium on the topic at the University of Witwatersrand.[71] He was also chair of the Alexandra Township Health Committee in the early 1940s where the University's Medical School indicated interest in establishing contraceptive services.[72] His involvement in the RWS may also have grown out of his research involvement in IQ testing, a popular activity with strong eugenic associations.[73] In 1925 Hoernlé supervised the testing of the IQs of 20,000 English and Afrikaans-speaking white children on the

Rand and in Pretoria as part of an assessment of the educational policy to instruct English- and Afrikaans-speaking children in their mother tongue.[74]

The Hoernlés joined the RWS in reaction against the rapid growth in the urban African population in interwar Johannesburg and the increasingly visible problem of black poverty and ill health. After 1932, the push of extreme rural poverty and pull of a growing number of manufacturing and other semi-skilled jobs (the result of South Africa's booming economy beginning in 1933) caused a rapid influx of Africans to the Rand. After South Africa abandoned the gold standard in December 1932, the mining industry flourished and this in turn engendered recovery and expansion in manufacturing industries. In the southern Transvaal, the African industrial labour force grew from 36,153 to 80,722 between 1932 and 1936.[75] Between 1928 and 1936, the number of African women alone almost quadrupled, rising from 29,000 to 107,000; the number of white women rose moderately by comparison in the same period, from 112,000 to 196,000.[76] (There were only 15,000 Coloured women on the Rand by 1936.[77]) This suggests that the Hoernlés' motive for delivering contraceptive services was in part to limit African population growth. In order to understand why two of the country's leading liberals would work towards this goal, it is necessary to place South African liberalism in context. According to Paul Rich, after the passage of the 1923 Urban Areas Act, which decreed that Africans were only allowed in urban areas in order to minister to white needs, the increasingly dominant political logic of segregationism defined the direction of South African liberal thought and action.[78] By the 1930s, a decade of intensifying impoverishment and exploitation of Africans in both rural and urban areas, the SAIRR (based in Johannesburg) offered only qualified criticism of the state for its increasingly draconian policies and demands to remove blacks from South African civil society. Instead, led by intellectuals like R. F. A. Hoernlé, the organisation sought to forge a compromise with government segregationist ideology.[79] (Indeed, Hoernlé in particular is seen by some to have been complicit in the elaboration of a pragmatic, 'liberal' version of segregation.[80]) It did so by developing a justification of segregation based on a relativist philosophy that emphasised 'cultural differences' between whites and Africans, arguing that contact between the two had become destructive to Africans who were now being cast in essentialist terms as a rural people. Indeed, the very term 'race relations' in its name signified the SAIRR's belief that innately different black and white cultures existed with inevitably conflicting interests. Targeting African, as well as Coloured and Asian, women for contraception, therefore, was probably an attempt to curb black population growth. By the 1930s, the central state was especially desirous of barricading the cities against unemployed African women. Historians have shown that African women, especially Basotho women, were regarded as a serious impediment to segregation and social order because of their success at illegally brewing and trading beer in shebeens located throughout slumyards and the African freehold neighbourhoods of Sophiatown, Newclare and Martindale.[81] Less well known is the state's serious concern about African women's fertility. Deborah Posel writes, 'Not only were women less likely than men to enter the formal urban labour market; their presence was equally "undesirable" for its effect on the size of the urban population, which expanded exponentially with each generation of children.'[82] In subscribing to the state's policy on segregation, the Hoernlés likely saw birth control as a way to contain the problem of the ever-growing urban black population.

The SAIRR's 'cultural relativist' research was heavily influenced by the emergent field of anthropology at the University of Witwatersrand where Winifred Hoernlé was a lecturer in social anthropology. This sheds additional light on to the Hoernlés' determination to establish birth control clinics as a means to stem black population growth. As Dubow points out, the founders of this discipline in South Africa believed it could be a source of applied knowledge from which solutions to 'race' problems could be derived.[83] A. R. Radcliffe-Brown, a co-founder of modern social anthropology, was based at the University of Cape Town in the early 1920s and claimed that new knowledge in this field was 'not merely of scientific or academic interest, but of *immense practical importance*'.[84] Not content simply to publish academic research, the Hoernlés sought a hands-on approach which would accommodate segregationist ideology while simultaneously assisting Africans.

There was undoubtedly a philanthropic dimension to these liberals' birth control activities. This was evidenced in the attitude of Winifred Hoernlé who was deeply troubled by the appalling conditions in the slumyards and squatter camps where newly arrived African workseekers settled. African misery, previously 'invisible' in the reserves, was now starkly evident to observers like herself. The African infant mortality rate in Johannesburg is estimated to have been almost ten times higher than for whites: in 1928, there were reportedly 705 deaths for every 1,000 black babies as compared to 78 for whites (though statistics for the black birth rate are notoriously unreliable[85]). Doubtless, one motive she had for distributing contraception was to stem this appalling loss of black child life. The impression that she joined the birth control movement out of concern for the health and welfare of black children is strengthened by the fact that in 1932, two years *before* joining the RWS, she was a founding member of the Joint Committee for Non-European Work established under the Johannesburg Child Welfare Society to coordinate welfare work among non-whites. Also, upon her resignation from the RWS in 1951 she took up the presidency of the Johannesburg Child Welfare Organisation and retained this position until her death in 1960.[86] As a result of Fantham's and Porter's departure for Canada in 1933, and the clinic's broader role in serving maternal health needs, the RWS became attractive to liberals with an evident desire to help Africans while simultaneously being compatible with segregationist policy. By 1934 the RWS had opened up space for members more interested in addressing poverty among all 'races' than in eliminating 'defective' whites.

From the middle of the decade onwards, the RWS forged fruitful institutional linkages with other social welfare agencies, the state (including MPs, the Departments of Public Health and Social Welfare) and the press.[87] Despite residual opposition from whites fearing 'depopulation' and ongoing condemnation from the Catholic Church, the tide of public opinion was turning in favour of the birth control movement. By 1939 the Children's Aid Society, Rand Aid Society and District Nurses Association were sending the RWS names of women deemed appropriate for birth control.[88] Official approbation was also forthcoming: at the organisation's 1939 annual general meeting, compliments were paid by Dr E. H. Cluver, Thornton's successor as Secretary for the Department of Public Health and Chief Medical Officer for the Union.[89] Also at that meeting, Dr Henry Gluckman expressed his appreciation of the RWS's work, supporting it on 'social grounds', and thanked Cluver for his 'wise support' of the organisation.[90] Such endorsements were accompanied by financial support. Beginning in 1934 the City Council of Johannesburg made annual grants to the RWS ranging in amounts from £100 to £200, in the hope that it would help to address the underlying cause of slum conditions which troubled municipal authorities by the early 1930s.[91] In 1933, Dr Milne, the Medical Officer of Health for Johannesburg, reported to the city's Public Health Committee 'that a white housing problem had been created by the influx of poor whites'.[92] And since 1924 the Johannesburg City Council had been attempting to relocate Africans from recently proclaimed 'white' neighbourhoods to newly built housing in African locations, but was unable to keep up with the number of Africans in-migrating to the city in search of work.[93] Birth control was likely regarded as an important aspect of attempts to contain the number of poor blacks and whites, and so in 1934 money was allocated to the RWS and in 1938 Milne joined the executive committee.[94] Most significantly, in 1938 the Department of Public Health began making annual grants of £1,000 to the national birth control coalition, of which the RWS received a substantial portion. In sum, by the late 1930s the RWS was perceived as a legitimate health and social welfare organisation by official and private agencies. Its growing public profile and credibility was symbolised most vividly by the decision to move the Central Clinic (as the Women's Welfare Centre was renamed in 1937) to the newly built Welfare House at 158 Fox Street in 1940 in order to facilitate cooperation with other social service agencies.[95]

Ironically, despite this success, the RWS did not sustain the level of popular support it had enjoyed at the start of the decade. There were fewer public lectures in the years after 1932, which in turn were attended by smaller audiences; this comes as no great surprise given that the results of birth control fell far short of the promised miracles of social and physical transformation. In short, the novelty of eugenics faded and popular interest in the RWS flagged as a result. Moreover, Fantham's and Porter's departure for Canada in early 1933 deprived the RWS of a spokesperson capable of effectively communicating eugenic ideas to a non-scientific audience. Most important, the end of the worst phase of South Africa's economic crisis by 1933 led to a decline in popular interest in eugenics as relative social stability and prosperity returned and 'poor whiteism' gradually but inexorably decreased. For the executive committee, which was continually distressed by the drop in popular interest, the RWS's rising reputation as a respectable, even mainstream, health and social welfare agency was crucial compensation for the loss of interest in its radical desire to transform the white 'race' through selective breeding and other controversial interventionist measures.

The RWS did continue to promote negative eugenics throughout the decade despite waning popular support. In 1936, a lecturer at an RWS event declared that birth control was insufficient to curb the spread of 'inferior elements'. Segregation and sterilisation of the 'feebleminded', drunkard and 'habitual criminal' were also necessary because from the point of view of the white race, 'some individuals must suffer for the benefit of all'.[96] As late as 1938, the RWS's statement of aims still placed negative eugenics at the top of its list: ensuring there would be fewer children from ill and 'shiftless' parents preceded the objectives of teaching 'mothers' to space their families for their own and their children's benefit, and promoting gynaecological care for mothers.[97] But by the end of the decade the executive committee's hardline conservative views, first espoused by Fantham and Porter and afterwards by Edward Mansfield, were marginalised as the liberals' urban welfare approach took hold.

Conclusion

The RWS was formed in 1930 in reaction against the social upheaval in Johannesburg wrought by the Great Depression. Led by H. B. Fantham and Annie Porter, the RWS was a eugenic organisation that advocated birth control as a 'scientific' solution to a wide range of social problems, including the proliferation of 'unfit' 'poor whites'. These indigent, predominantly Afrikaans-speakers who flooded to urban centres in search of employment, were despised by the mainly English-speaking professionals who founded the RWS and who resented having to subsidise state social welfare measures created to support them. Moreover, 'poor whites' were held to threaten white supremacy in South Africa because they were feared to be 'breeding' uncontrollably – among themselves as well as with blacks – thereby diluting the quality of the white 'race'.

The RWS established the first birth control clinic in South Africa and between 1932 and 1937 it opened an additional four. The clinics for African, Coloured and Asian women were established by the liberal members of the RWS's executive committee who joined after Fantham and Porter had departed South Africa, and after the mandate for birth control had been altered to include a broader focus on improving maternal health at the instigation of the Department of Public Health. Winifred Hoernlé was particularly important in this shift in focus for she helped to modify the RWS into a mainstream, respectable health and social welfare agency through which newly enfranchised middle-class white women like herself could help develop and implement public policy. Conservative members of the executive committee remained active in the RWS until the end of the 1930s and tried repeatedly to return the RWS to its original, negative eugenic roots. They were thwarted in their attempts to do so, in large part because the end of the Depression led to an evaporation of popular interest in eugenics.

However, the RWS achieved significant results. It opened up space in popular and official circles for discussion about birth control; raised greater awareness about maternal ill health; and attained legitimacy and financial support for medicalised contraceptive services from social welfare agencies and various levels of the state. Finally, but by no means least important, the RWS provided contraception and gynaecological care to thousands of women from all 'races' during the

THE RACE WELFARE SOCIETY

1930s. The organisation remained active until the late 1950s and has continued to exist in different forms and under various names into the present. Indeed, the role and legacy of the birth control movement in the development of the central state's 'population control' programme aimed at blacks during the apartheid era has yet to be understood.

Notes

1 I would like to thank Alan Jeeves and Saul Dubow for their very valuable comments on an earlier draft of this chapter. Thanks also to Dr Elin Hammar for sharing the RWS documents with me, and for her warm hospitality during my stay in Johannesburg in 1997.
2 H. B. Fantham, 'Eugenics', *Child Welfare*, 9:7 (1930), 5.
3 A. Porter, 'Eugenics from a woman's point of view', *Child Welfare*, 10:2 (1931), 3.
4 Historical Papers, University of Witwatersrand, Johannesburg, Race Welfare Society Documents (hereafter RWS), minutes, 26 June 1930.
5 National Archives of South Africa, Pretoria (hereafter NASA), SAB GES 2281 85/38, vol. 1, letter by Joint Secretaries of the RWS, 29 November 1930.
6 Fantham, 'Eugenics', p. 6.
7 Charles Webster, 'Introduction,' in Charles Webster (ed.), *Biology, Medicine and Society 1840–1940* (Cambridge, 1981), pp. 1–13. The study of eugenics movements in other contexts is growing rapidly. A good place to begin is Mark Adams (ed.), *The Wellborn Science: Eugenics in Germany, France, Brazil, and Russia* (New York, 1990).
8 Mark Adams, 'Toward a comparative history of eugenics', in *The Wellborn Science*, pp. 217–26.
9 Saul Dubow, *Scientific Racism in Modern South Africa* (Cambridge, 1995); Susanne Klausen, ' "For the Sake of the Race": Eugenic discourses of feeblemindedness and motherhood in the *South African Medical Record*, 1903–26', *Journal of Southern African Studies*, 10:1 (1997), 27–50.
10 Helen Bradford, 'Herbs, knives and plastic: 150 years of abortion in South Africa', in T. Meade and M. Walker (eds), *Science, Medicine and Colonial Imperialism* (London, 1991); Joyce Newton-Thompson, 'Birth control clinics in the Western Cape, c. 1932 to c. 1974: A history', BA thesis, University of Cape Town, 1992; Catherine Burns, 'Reproductive labors: The politics of women's health in South Africa, 1900–1960', Ph.D. dissertation, Northwestern University, 1995.
11 RWS, 14 August 1930.
12 RWS, 31 December 1934.
13 In South Africa during the 1930s and after, non-whites were often viewed *en masse* in oppositional terms to whites, therefore the term 'blacks' is used in this chapter to encompass all non-white population groups (Africans, Coloureds and Asians).
14 Susanne Klausen, 'The formation of a national birth-control movement and the establishment of contraceptive services in South Africa, 1930–1939', Ph.D. Dissertation, Queen's University at Kingston, 1999.
15 In February 1997 I contacted Dr Elin Hammar who had worked for the RWS beginning in 1950, in the hope that she could help me locate the RWS's archive. Fortunately, she had the documents herself, having saved them from destruction during an office relocation in the 1970s. Dr Hammar was kind enough to lend me the material and afterwards agreed to donate it to the William Cullen Library at the University of Witwatersrand.
16 'Loss to the Rand: Professor Fantham for Canada', *Star*, 21 September 1932.
17 'Honour for Prof. Fantham – The South Africa Medal', *Star*, 10 June 1931.
18 Wellcome Institute for the History of Medicine, Contemporary Archives, London

[183]

(hereafter WIHM), SA/EUG E.22, ESSA pamphlet by A. J. T. Janse, 'The Eugenics Society of South Africa,' n.d. As a study group, the Eugenics Society of South Africa (hereafter ESSA) failed to attract popular support and it petered out in November 1931. The RWS succeeded where the ESSA failed because of its practical programme of establishing birth control clinics.

19 Fantham and Porter's articles include: 'Notes on some cases of racial admixture in South Africa', *South African Journal of Science* (hereafter *SAJS*), 24 (1927); 'Some further cases of physical inheritance and racial admixture observed in South Africa', *SAJS*, 27 (1930); 'Inheritance of stature through mate-selection', *Journal of Heredity* (September 1935); 'Remarks on a "family" showing shortness, illegitimacy and simplemindedness', *Eugenical News* (March–April 1936). Articles Fantham authored on his own are listed in Dubow, *Scientific Racism*, pp. 132–7.
20 Marais Malan, *The Quest for Health: The South African Institute for Medical Research* (Johannesburg, 1988), p. 35.
21 Ibid., p. 37.
22 Fantham, 'Eugenics', p. 6.
23 Ibid., p. 3, and Porter, 'Eugenics from a woman's point of view', p. 4.
24 Fantham, 'Eugenics', p. 7.
25 Dan O'Meara, *Volkskapitalisme: Class, capital and ideology in the development of Afrikaner nationalism 1934–1948* (Cambridge, 1983), pp. 36–7.
26 Elsabe Brink, 'The Afrikaner women of the Garment Worker's Union, 1918–1938', MA thesis, University of the Witwatersrand, 1986, p. 45.
27 Jonathan Hyslop, 'White working-class women and the invention of apartheid: "Purified" Afrikaner nationalist agitation for legislation against "mixed" marriages, 1934-9', *Journal of African History*, 36 (1995), 62.
28 WIHM, SA/EUG E. 22, ESSA pamphlet by A. J. T. Janse, 'The Eugenics Society of South Africa', n.d.
29 'Henry Britten', in Ken Donaldson (ed.), *South African Who's Who* (Cape Town, 1937 and 1944 edn). The Coronation Medal was awarded to leading social welfare reformers, philanthropists and other notables on the occasion of the coronation of King George VI.
30 Susan Parnell, 'Slums, segregation and poor whites in Johannesburg, 1920-1934', in Robert Morrell (ed.), *White But Poor: Essays on the history of the poor whites in southern Africa* (Pretoria, 1992), p. 22.
31 Hardy was a partner in the accounting firm Howard Pim and Hardy. Pim was among the first theorists of segregation to outline a plan for the reserve system for Africans in South Africa, but by 1930 he was a prominent liberal critic of segregation. See Saul Dubow, *Racial Segregation and the Origins of Apartheid in South Africa, 1919–36* (London, 1989), pp. 23–5.
32 'Mr J. L. Hardy, O.B.E.', *Rand Daily Mail*, 26 November 1941.
33 RWS, 21 September 1932.
34 RWS, 1 September 1931.
35 RWS, 30 June 1932. Also, Dr Berry of the executive committee took ten District Surgeons to visit the RWS's first birth control clinic in early 1932. Afterwards, many requested supplies of contraceptives. RWS, 25 February 1932.
36 For example, Dr I. Block spoke in favour of birth control at the South African Medical Council's annual meeting in 1934 and his speech was reprinted in the medical press. I. Block, 'Observations from the work of a birth control clinic', *South African Medical Journal* (14 July 1934), 490–2.
37 Throughout the 1920s the Dutch Reformed Church Synods had regularly passed resolutions condemning birth control as immoral. E. G. Malherbe, *Never A Dull Moment* (Cape Town, 1981), p. 130. The Church was persuaded to revisit the issue after being lobbied in favour of birth control by two investigators for the Carnegie Commission of Inquiry Into the Poor White Problem (1929–32), Malherbe and M. E. Rothmann, and by white Afrikaans-speaking women's welfare organisations. In 1934, a commission on birth control found that contraceptives were appropriate in cases where a woman's life was in danger, where she had already given birth to two

consecutively born children who were either blind or disabled, or where she already had children who were underfed. Marie Stopes, *Birth Control News*, 13:2 (June 1934), 26. By 1936, a birth control activist from Pretoria reported that when the issue of birth control had been brought before the Transvaal Synod of the Dutch Reformed Church, 'that body did not in any way express condemnation of birth control'. Planned Parenthood Association of South Africa, Johannesburg, 'Minutes of the Second Meeting of the South African National Council for Birth Control', 6–7 October, pp. 6–7. The Catholic Church remained a vociferous critic of municipal and national governments' support for birth control.

38 See her article praising the Cape Town birth control clinic, in *Die Burger*, 3 August 1933; and 'The mother and daughter of the poor family', *Report of the Carnegie Commission of Inquiry Into the Poor White Problem*, vol. V (Stellenbosch, 1932).
39 For discussion about opposition to birth control from Afrikaner nationalists and whites in general see Klausen, 'The formation of a national birth-control movement', 1999.
40 Block, 'Observations from the work of a birth control clinic,' p. 492.
41 *Ibid*.
42 RWS, 27 August 1930.
43 Two female physicians who worked in Cape Town's birth control clinics in the 1930s stated this was a prominent reason. Interviews with Dr Patricia Massey, Cape Town, April 1997, and Dr Dorothea Douglas-Henry, Vishoek, April 1997.
44 NASA, SAB GES 2281 85/38, vol. 1, RWS document, 'Objects of the Society', n.d.
45 RWS, 27 August 1931, and RWS, 21 September 1932.
46 Regarding public events and public attendance:
1930: 18 September, 'The economic aspects of birth control', by Miss Pollack and Dr Bernstein, 102 people attended; 30 October, 'Ethics of birth control', by O. C. Jensen, 74 people attended.
1931: 26 February, 'Inheritance of mental defect', by Dr A. M. Moll and 'The economic aspects of mental defectiveness', by Miss Pollack, about 70 people attended; 26 March, 'The social worker and eugenics', by Dr Maria Te Water, 47 people attended; 28 April, 'Why South Africa needs birth control', by George Hills MP, 70 people attended; 25 June, 'Eugenics versus disease, poverty and war', by Dr Annie Porter and G. H. Shawe, 35 people attended; 27 August, 'Glands and personality', by H. B. Fantham, 318 people attended; 29 October, 'Points connected with heredity in South Africa', by Dr Porter, 13 people attended.
1932: 18 March, 'Birth control, wages and workers', by George Hills MP, Dr F. Berry, Dr A. M. Moll and Henry Britten, about 50 people attended; 21 September, 'Biology and civilisation', by H. B. Fantham, 220 people attended.
Symposium: 26 May 1931 the RWS and Mental Hygiene Society co-sponsored 'The poor white question', and about 220 people attended. Dr Morris Cohen was the Chairman and speakers included Drs Frankel, Malherbe, Te Water and Block, and Mr Radloff.
For reportage in prominent newspapers see: 'Future welfare of the race. Serious problem of the feeble-minded. Sterilisation or segregation?' *Rand Daily Mail*, 15 August 1930; 'Improvement of the race', *Star*, 15 August 1930; 'Birth control a necessity. Outgrowing world's food supply', *Rand Daily Mail*, 25 August 1930; 'Birth control clinics urged. "A duty to the nation." Support from well-known speakers', *Rand Daily Mail*, 29 April 1931; '19 Children on 4/- a day. Poor man's huge family. Need of birth control clinics stressed', *Star*, 29 April 1931; 'Expert views on poor whites. A valuable symposium', *Rand Daily Mail*, 28 May 1931; 'Problem of poor whiteism. Habits can be changed. Questions reviewed from five angles', *Star*, 28 May 1931.
47 White women acquired the franchise in 1930.
48 See Morrell, *White But Poor*.
49 *Report of the Carnegie Commission*, vol. IV, p. vii.
50 O'Meara, *Volkskapitalisme*, p. 32.
51 Cited in Deborah Gaitskell, '"Getting close to the hearts of mothers": medical

missionaries among African women and children in Johannesburg between the wars', in Valerie Fildes, Lara Marks and Hilary Marland (eds), *Women and Children First: International maternal and infant welfare 1870–1945* (London, 1992), p. 186.
52 Fantham, 'Eugenics', p. 3.
53 *Ibid.*, pp. 5–7.
54 *Ibid.*, p. 5.
55 'Future welfare of the race. Serious problem of the feeble-minded. Sterilisation or segregation?' *Rand Daily Mail*, 15 August 1930.
56 RWS, 27 November 1930.
57 NASA, SAB GES 2281 85/38, vol. 1, fundraising letter from EC, 29 November 1930.
58 NASA, SAB GES 2281 85/38, vol. 1, letter by Joint Secretaries of the RWS, 17 February 1931.
59 See, for example, *Debates of the House of Assembly*, vol. 14 (17 January to 4 April 1930), 24 January, p. 77. As a result of growing alarm over white maternal mortality the Department of Public Health established a rural nursing programme in 1935.
60 NASA, SAB GES 2281 85/38, vol. 1, letter from the Secretary for the DPH to the EC, 11 March 1931.
61 'Improvement of the race', *Star*, 15 August 1930.
62 '19 children on 4/– a day. Poor man's huge family. Need of birth control clinics stressed', *Star*, 29 April 1931.
63 *Ibid.*
64 RWS, 4 December 1936.
65 Elsewhere I have examined Thornton's crucial impact on the development of the birth control movement in South Africa. See, 'The South African state, white maternal mortality and birth control, 1931–1939', unpublished paper.
66 NASA, SAB GES 2281 85/38, vol. 1, letter from EC to the Secretary for the DPH, 27 May 1931, p. 2.
67 *Ibid.*
68 Sallie Woodrow, 'The start of the family planning movement in Cape Town', unpublished MS, September 1972.
69 'Dr Hoernlé dies. A tireless social worker', *Rand Daily Mail*, 18 March 1960; 'Dr Hoernlé dies', *Star*, 17 March 1960.
70 John Laird, 'R. F. Alfred Hoernlé', *Notes*, p. 286. Archive of the University of Witwatersrand.
71 NASA, SAB GES 2281 85/38, vol. 2, pamphlet, 'Race welfare in South Africa', Pretoria, 1940, which contains E. H. Cluver's address to the Race Welfare Society, p. 1.
72 RWS, 1 December 1939.
73 'R. F. A. Hoernlé dead', *Star*, 21 September 1943.
74 R. F. A. Hoernlé, 'Wits University Group Test of mental ability, 1925', Pretoria, 1926; 'Intelligence tests', *Star*, 2 August 1926.
75 Andre Proctor, 'Class struggle, segregation and the city: A history of Sophiatown, 1905–40', in Belinda Bozzoli (ed.), *Labour, Townships and Protest: Studies in the social history of the Witwatersrand* (Johannesburg, 1979), p. 65.
76 Hyslop, 'White working-class women and the invention of apartheid', p. 62.
77 *Ibid.*
78 Paul Rich, *White Power and the Liberal Conscience: Racial segregation and South African liberalism* (Manchester, 1984), pp. 6–7.
79 The summary of the SAIRR is based on *ibid.*
80 Martin Legassick, 'Race, industrialization and social change in South Africa', *African Affairs*, 75:299 (1976), 224–39.
81 Phillip Bonner, ' "Desirable or undesirable Basotho women?" Liquor, prostitution and the migration of Basotho women to the Rand, 1920–1945', in Cherryl Walker (ed.), *Women and Gender in Southern Africa to 1945* (London, 1990), pp. 221–50; Cherryl Walker, 'Gender and the development of the migrant labour system, c. 1850–1930', in *ibid.*, especially pp. 180–7; Julia Wells, *We Now Demand: The history of women's resistance against the pass laws in South Africa* (Johannesburg,

THE RACE WELFARE SOCIETY

1993); and Cherryl Walker, *Women and Resistance in South Africa* (2nd edn, Cape Town, 1991); Proctor, 'Class struggle, segregation and the city'.
82 Deborah Posel, 'State, power and gender: Conflict over the registration of African customary marriage in South Africa, c. 1910–1970', *Journal of Historical Sociology*, 8:3 (1995), 230.
83 Dubow, *Racial Segregation*, p. 34.
84 Cited in *ibid*.
85 Deborah Gaitskell, 'Housewives, maids or mothers: Some contradictions of domesticity for Christian women in Johannesburg, 1903–39', *Journal of African History*, 24 (1983), 250, and Gaitskell, 'Getting close to the hearts of mothers', p. 199.
86 *Standard Encyclopedia of Southern Africa* (London, 1972), p. 551.
87 'Society for Race Welfare. Expansion claimed in Annual Report', *Star*, 30 October 1936; 'Transmission of disease. Address to Race Welfare Society', *Star*, 4 November 1936; 'Survival of degenerates. Eugenics combats race menace', *Rand Daily Mail*, 4 November 1936; 'Practical clinic work. The birth control movement', *Star*, 1 August 1937; 'Birth control in city. Race welfare meeting', *Star*, 20 November 1937; 'Attendance at race welfare clinics lower', *Rand Daily Mail*, 20 November 1937.
88 RWS, 9 June 1939.
89 RWS, 13 October 1939.
90 *Ibid*.
91 Parnell, 'Slums, segregation and poor whites,' p. 123.
92 *Ibid*.
93 Proctor, 'Class struggle, segregation and the city', p. 59.
94 RWS, 21 July 1938.
95 RWS, 20 December 1938.
96 'Transmission of disease. Address to race welfare society', *Star*, 4 November 1936. Also on 29 October 1941 the RWS and National Council for Mental Hygiene co-sponsored a symposium on 'Voluntary sterilization'.
97 RWS, 30 June 1938.

CHAPTER EIGHT

Doctors and the state: George Gale and South Africa's experiment in social medicine[1]

Shula Marks

The conception of social medicine

What constitutes cutting-edge scientific research in a country like South Africa? This question is of more than usual interest in the contemporary Republic as competing notions of priority, relevance and universal standards of research vie with one another in the marketplace of ideas, both within the medical profession and between the profession and government. For a small group of medical practitioners in the 1940s and 1950s the answer seemed to lie in developing notions of social medicine through family and community health centres. In this field, for just over a decade in the mid-century, South Africa was widely acknowledged as being in the forefront of international progressive thought, its distinctive social conditions and developed medical practice making possible an experiment in social medicine with far-reaching implications. Not only did Dr John B. Grant of the Rockefeller Foundation consider that there was 'nothing in the whole world more advanced' in concept and practice than the Pholela health centre in Natal, pioneered and run by Drs Sidney and Emily Kark – no mean praise from the man who had himself pioneered health services in China and India;[2] the publication of the 1944 National Health Services Commission report (the NHSC or Gluckman report) also led Malcolm Macdonald at the Ministry of Health in London to exclaim, 'This is a report that shows us what we should be doing!'[3] The curriculum of the newly formed medical school at the University of Natal embodied principles which are being rediscovered in the late 1990s, while the Institute of Family and Community Health established to train health workers for the health centres under Sidney Kark's direction was equally noteworthy. Moreover, with the dispersal of South Africa's most progressive doctors in the 1950s their ideas spread to many parts of the world.

The conception of social medicine was of course neither new nor unique to South Africa. As Dorothy Porter has shown, in Britain 'the original conception was built upon a collection of beliefs about the nature of science and medicine which were shared by various branches of the [medical] profession... with diverse social values'. With the appointment of John Ryle to the first chair in social medicine in Oxford in 1942, 'the synthesis of ideas that created the discipline... were integrated into a specifically left-wing philosophy of social reform'. Ryle's thinking was influenced by the scientific humanism of his day, Smuts's holism and the Soviet model of planning.[4] All these influences were absorbed, as we shall see, in various ways by medical reformers in South Africa. What differentiated the South African experiment, however, was the evolution of a new set of practices around primary health care imaginatively adapted to local experience and, for a brief period, their incorporation into government health planning. Thus, while health centres were not new, they were given a particular significance in the Gluckman report as the building blocks for a national health service, in which they were intended to provide 'comprehensive, fully integrated medical care' through a single team, responsible for preventive and promotive (educational) as well as curative services.[5]

Crucial in these achievements was a small group of progressive doctors, most notably Sidney and Emily Kark, whose endeavours have been internationally celebrated. Thus, Dr Mervyn Susser, Emeritus Professor of Public Health at Columbia University and former editor of the *American Journal of Public Health*, has recently argued that the WHO's milestone Alma Ata Declaration advocating primary health care was 'an embodiment and a triumph of Karkian principles'.[6] Less well known but quite crucial was the support the Karks received at the time from key individuals within the Department of Health, such as Eustace H. Cluver as Secretary of Health (1938–40), Harry S. Gear (Deputy Chief Medical Officer from 1939, and in the 1950s an Assistant Director General of WHO) and George Gale, who was appointed Chief Medical Officer and Secretary for Health in 1946 by Dr Henry Gluckman, South Africa's only progressive Minister of Health before 1994. Without their assistance and the recommendations of the Gluckman Commission, the Karks' pioneering achievements in South Africa would have been if not impossible, at least far more limited, as they have been the first to acknowledge.[7]

George Gale and social medicine in South Africa

Of these influential individuals, George Gale is of particular importance. In the Department of Public Health from 1939 until 1952, Gale was, according to Sidney Kark, 'the brain behind the Gluckman report', much of which he almost certainly wrote.[8] He was also the bridge between Kark and Gluckman.[9] At a time when Kark's ideas were by no means accepted by the majority of medical practitioners or even by all members of the Department, his constant vigilance and reasoned defence of the health centres as well as of the Institute of Family and Community Health, which was designed to train health workers for the centres and undertake research into social medicine, was frequently decisive. As Emily Kark wrote to George Gale's widow, Audrey, shortly after her husband's death in 1979, 'By its very nature the Polela Health Centre challenged conservative government opinion and but for the valiant and persistent efforts of George Gale it might earlier have been adversely affected by the establishment.'[10] Gale's concern for the health needs of rural Africans and his support in the Department of Health was crucial in ensuring the survival of at least some of the health centres even after the Nationalists came to power in 1948. As he himself saw it in 1950, he was 'lucky' that he had 'been able to persuade political masters to do as much as they have done'.[11]

Moreover, it was Gale who ensured that the new Medical School in Durban, 'for Africa' as he thought of it, was placed in a major city and not in a 'reserve'. As its first Dean, he insisted that it be on a par with any other medical school in the country and he ensured that the Institute of Community and Family Health at Clairwood was consolidated as part of the Department of Social, Preventive and Family Medicine in the Faculty of Medicine. With Kark as its first professor, and with encouragement and funding from the Rockefeller Foundation, Gale integrated its work into the School's clinical training.[12] On his retirement in the 1960s, he took many of these ideas to the medical schools of Thailand and Malaysia where he was visiting professor supported by the WHO, and helped set up curricula in public health and preventive medicine.

Gale was 'a complex and contradictory figure... caught up and influenced by the socio-political historical context of segregationist and later apartheid South Africa'; despite his best efforts, the medical school he headed was deeply embedded in the racist norms of the white South African state.[13] Nor does he fit comfortably within either a radical or a conservative paradigm of South African politics. Thus, in 1974, at the very time that he was arguing publicly against the expulsion of the South African Medical Association from international medical organisations, and defending South African doctors against charges of racism, we find him writing privately to his friend, the missionary-doctor Douglas Aitken, who was highly indignant at assistance provided by the World Council of Churches for the African National Congress:

> Is the use of force ever morally justifiable? The initial reaction in this country to the W.C.C. action was not, in the main, as in S. Africa, one of anger but rather of concern to know why an organisation covering so large a part of Christendom felt justified in giving material support to guerrillas in Southern Africa; and almost at once the question was put: Who really are the terrorists in Africa – the Whites or the Blacks? ... The feeling around the world now is, and surely with justification, that nothing is ever likely to soften the hearts of enough White S. Africans to bring about fundamental change in their Government's racist policy. According to Holy Writ, after Pharaoh had refused Moses' plea to 'let my people go', God Himself terrorised him into doing so.[14]

By this time, Gale's views had probably been influenced by his five-year sojourn in Uganda in the run-up to independence, although he had already made contact with black medical students from South Africa in Edinburgh in the mid-1930s and showed a rare understanding of their perspective. Anxious to counter allegations that African medical students educated overseas would 'become infected with subversive political and social ideas', he retorted: 'As if students cannot read! – or as if Natives need to go outside of their country to become conscious of political and social inequalities *within* their country.' On the contrary, he maintained,

> Far from being political firebrands, among them are some of the sanest and most conciliatory leaders of the Abantu ... I was interested to note that they had themselves realised, through their contact with the poorer sections of Edinburgh population, that many of the problems which in South Africa they had thought were the outcome of racial oppression existed also in Britain, and were in essence economic.[15]

At Fort Hare in the late 1930s Gale had close contact with his African students and recorded their frustrations, while in the early 1940s he noted being impressed by Dr A. B. Xuma.[16] It appears that African physicians were equally impressed with Gale, for he was assured by Dr Moroka, the then President of the ANC, and others, that African opposition to the new medical school would end because his becoming Dean in 1952 was 'a guarantee that it will not prove to be another Fort Hare type of hoax'.[17]

Nevertheless, as a man who held a top state appointment, Gale continued to believe in the efficacy of working within the system, 'having great faith in his power of persuasion', and thought he could convince ministers by rational argument even after the Nationalist victory of 1948.[18] Through George Gale's career it is thus possible to explore the possibilities for, and constraints upon, doctors attempting medical reform from 'within the system' in South Africa, and to recapture some of the history of social medicine in the mid-twentieth century. In fostering social medicine and a new form of medical education, his thinking was undoubtedly in advance of that of the vast majority of his medical contemporaries in South Africa and abroad.

Like his fellow-advocates of social medicine, Gale had little doubt about the socio-political causes of much of South Africa's burden of ill health and explicitly advocated socio-political means of preventing and curing it, even if in the final analysis he was powerless to implement the broader policies he believed necessary. According to Roy Gale, his father had 'ranted against the effects of the Native Recruiting Corporation [which recruited rural Africans for the mines] on both health and society ... well back into his missionary days'.[19] As a mission doctor, he neither romanticised 'healthy reserves', nor blamed 'the victims' for their susceptibility to disease.[20] His evidence before the Mine Wages Commission of 1943 – the so-called Lansdowne Commission – makes this abundantly clear:

> The system of migratory labour [he claimed] was harmful to social, economic and health conditions in the reserves. It handicapped the development of the reserves because it drained manpower ...
>
> Mine wages plus reserve income did not provide the native with an adequate income. Many natives in reserves lived below the breadline and the fact that the fittest of them increased in weight by anything up to one stone on the mines showed that they were badly nourished.
>
> ... Natives on the mines should be given a living wage so that they could live with their families without assistance from the reserves. This might result in the closing down of some mines and then fewer natives would have to be accommodated in Johannesburg and more would be available to develop the reserves. Revenue from the mines would probably decrease if this was done, and the standard of living of Europeans would be lowered, but the long-term benefits to the whole country would outweigh the disadvantages.[21]

Gale was equally forthright a few years later when, already Secretary of Health, he declared that tuberculosis and venereal disease could not be cured simply by building additional hospitals and clinics: 'The overall campaigns must include radical changes in the socio-economic system under which non-European labour is recruited, housed and fed.'[22]

George Gale had a very distinct notion of the responsibility of society for the health of its citizens; aware of the social causes of much ill health, he also recognised the social means necessary for preventing and curing it. As a result, he appears initially to have used the terms 'social' and 'socialized' interchangeably.[23] By the late 1940s, however, he was concerned to dissociate the two words, perhaps because of the rising tide of anti-communism both in South Africa with the advent of the National Party in government (when he was still a government servant) and more widely with the onset of the Cold War. As he was only too aware, by the late 1940s social medicine was reputed to be 'leftist' and, according to a somewhat hostile observer from the Rockefeller Foundation, this fact was 'not particularly beneficial to the health center development under Kark'.[24]

At the same time, Gale was also anxious to distance himself from those who believed that 'social medicine' denoted purely 'preventive medicine' to the exclusion of curative health; he was at some pains to define it as 'that practice of medicine in every department of medicine, which takes cognizance of the social as well as the purely physical factors in the aetiology of illhealth and in the promotion and preservation of good health'.[25] The definition captures much of the philosophy behind the Gluckman Report and the development of family and community medicine by the Karks.

If on the social and political causes of disease Gale's views were at variance with most of his peers, he undoubtedly shared with contemporary 'men of good will' 'an optimistic faith' in the possibilities of science and medicine in opening the way to 'a more humane, healthy and enlightened society'.[26] Like most missionary doctors – and indeed non-missionary practitioners – Gale was certainly impatient of what he saw as 'superstition' and the obstacle it posed to health. Before the Gluckman Commission, he insisted that 'ignorance, was a close rival to, if it did not actually exceed, poverty as a cause of disease among the Native population'.[27] Yet he was also capable of a degree of cultural relativism which was perhaps more unusual for his time, profession and place: he clearly did not think ignorance was confined to the black population, and he was alert to the alternative and sometimes more effective approaches that African healers took to the treatment of disease.

As Dean of Durban Medical School Gale was insistent that students be taught anthropology and sociology; Hilda and Leo Kuper, already eminent in these fields, were brought in to provide introductory courses. Throughout his career, despite an evident paternalism (perhaps an occupational hazard in a man who was both a missionary and a public health doctor), Gale was sensitive to the need to work with Africans. (This was also true of the Karks and their immediate colleagues who were seeking ways to liaise with community associations as early as 1942 and were often frustrated by the apparent absence of such organisations.)[28]

George Gale's background

Gale's commitment to what we would now term primary health care was rooted in his personal history. The son of missionaries, Gale grew up in northern Zululand where he learnt to speak fluent Zulu. He was a brilliant student and took his M.Sc. under John William Bews, Professor of Botany in Natal and the first Principal of the university. Bews's holistic ideas of human ecology (derived from Smuts who in fact wrote the introduction to his *Life as Whole*) remained with Gale throughout his life.[29] Assisted by illustrious scholarships and prizes, Gale trained as a doctor in Edinburgh, where he 'resided in a Slum Settlement undertaking social, religious and medical work' and where he was inspired by the public health advocate, Professor P. S. Lelean. In 1935 he also qualified for the Diploma in Public Health there.[30]

For Gale, however, medicine was not simply a profession: it was a vocation, part of the Christian mission he imbibed from his earliest years; indeed, according to Roy Gale, his entire life was inspired by a deep sense of religion.[31] From his student days in Natal and Edinburgh (where he devoted much of his leisure time to evangelistic activities) to his devotion while professor at Makerere to the Mengo Hospital Board (which he gave 'spiritual' as well professional leadership) his life was directed by Christian purpose.[32] As Reg Pearce, his brother-in-law and university contemporary, wrote in 1938 to the then Minister of Education, Health and Welfare, Jan Hofmeyr, George Gale was a 'man with ideas who is prepared to sacrifice something for those ideas, and who has the necessary energy to achieve results ... if ever there was a man with a single purpose, imbued with a strong missionary ideal, it is Gale'.[33]

In 1928, having completed his medical training, Gale returned to South Africa to restore the Gordon Memorial Mission (first at Pomeroy, later at Tugela Ferry) with the money for medical missions donated by Lord Maclay, a Glasgow shipowner. He practised medicine among rural Zulu for eight years before joining the staff at Fort Hare, the African college of further education in the eastern Cape, in August 1936. Gale's remit was to launch a new scheme to train African medical aids as rural paraprofessionals.

In its origins the Fort Hare scheme was designed to give a five-year training to candidates who had achieved a Junior Certificate level of education. In the event, the Fort Hare authorities insisted on accepting only matriculants (who had a further two years of schooling) for a five-year course, including a year's internship at McCord Hospital in Durban. As relatively few Africans matriculated in the 1930s, only a small number of aids were trained (twenty to twenty-five before 1945). To their intense frustration most found that, although their training was only a year short of the full medical degree, they were confined to the humblest of tasks. The scheme failed to provide the large cadre of rural health workers envisaged by the Department of Health, and Gale, concerned with its legal anomalies and sympathetic to African dissatisfaction, rapidly found himself out of tune with his Fort Hare colleagues. By April 1937, he had resigned in protest.[34]

For a short period in late 1937 and 1938 Gale worked as Assistant M. O. H. in Pietermaritzburg and Benoni, hoping for an appointment in the Union's Department of Public Health.[35] Still deeply concerned with the problems of rural health and indignant over what he saw as the travesty of the Fort Hare scheme, he wrote a pamphlet, *A Suggested Approach to the Health Needs of the Native Rural Areas of South Africa*, which he published at his own expense in 1938. It was this pamphlet that led the Secretary of Public Health, Eustace Cluver, to invite Gale to join his department and help develop rural health services.[36]

The origins of health centres and community-oriented primary health care in South Africa

Through the 1930s there had been rising concern with African ill health in both town and countryside as South Africa's cheap labour policies based on migrant labour and rural impoverishment took their toll in escalating TB rates, malnutrition and venereal disease: according to a survey conducted by the Karks in the late 1930s 23 per cent of children in rural and urban areas had congenital syphilis. Indeed, the training of medical aids was part of the state's initial if inadequate response. The rising number of Africans in urban hospitals was the visible tip of an iceberg of ill health – and one that increasingly worried provincial authorities responsible for the hospital services and reluctant to pay the cost.

It was in this context that Cluver and Harry Gear decided to set up three rural health 'units', and invited Sidney and Emily Kark to establish one such centre at Pholela.[37] (In the event, because of the war, the second centre was only set up at Bushbuckridge in the north-eastern Transvaal in 1943, although a Health Centre was established at Umtata by voluntary effort in 1941 and taken over by the Department in 1943.)[38] At about the same time the Karks also read Gale's pamphlet and met the author, who began to pass them literature on rural health needs; in 1938–39 they were meeting two to four times a week.[39] They continued to correspond after the Karks went to Pholela, and Gale often visited the centre, proffering advice and helping to resolve local problems.

The intellectual background and formation of the Karks and Gale could not have been more different: the Karks from East European Jewish and Zionist families, George Gale of Scottish missionary background. Nevertheless, a practical humanitarian socialism brought them together.[40] With his rural mission experience Gale had valuable insights to bring to the entire conceptualisation of the health centre scheme. As early as 1938 he was formulating ideas which were of relevance to the health centre idea.[41] At least as important was his strategic location at the heart of the health bureaucracy from 1940.

Two ironies should be noted here. The attraction of the health centre idea for the state was initially because it was seen as a low-cost way of dealing with rural health problems at a time when Africans were coming into hospitals in unprecedented numbers. Cluver announced this quite explicitly in a letter designed to appeal to the Secretary of Finance early in 1939: 'In order to control expenditure on Native hospitalisation and to reduce the demand in a manner which will ultimately result in a considerable saving ... a scheme has been evolved to establish inexpensive clinics for the early treatment of disease among Natives.'[42] Genuinely concerned with the rising tide of ill health among the African population, Cluver may have couched his letter in terms calculated to appeal to the treasury. As this text and others demonstrate, however, the main concern for many in the Department and for the provincial authorities was the cost of African hospitalisation, and this lay behind the establishment of the first two health centres.

In the same way – and this is the second irony – Gale's attractiveness to the Department of Public Health may have been his preparedness to work within the bounds of contemporary white prejudice. Thus his *A Suggested Approach* is in many ways a curiously contradictory text, where for every two progressive steps forward he took one step back. His ideas were often trapped in the segregationist discourse of his day.[43] It is almost impossible to convey these ambiguities without analysing the pamphlet line by line, if not word for word. Here a self-mocking passage on the relationship between African and 'scientific' medicine must suffice. On the one hand, Gale argued, traditional healers often established 'a bond of sympathy and confidence' with their patients which was 'as important a factor towards recovery as the remedies used'. As a result, Africans refused to abandon their 'superstitious beliefs' 'as we in our infinite superiority and condescension assume they must or should, just because we pooh-pooh them'. Scientific scorn simply persuaded the African to doubt the efficacy of the western doctor compared with the 'inyanga'

> who takes account of *all* the disease-producing factors and tackles them with a will. In fact, the Native who consults a European doctor is in much the same position as a civilised person who, having failed to get relief from orthodox practitioners, decides to try a quack who he knows does not accept orthodox ideas of current medical practice, but nevertheless achieves remarkable cures from time to time![44]

On the other hand, despite Gale's apparent relativism, and belief in holistic medicine, his bottom line remained that 'the [rural] African lives in a world irreconcilable to scientific theory'.[45] This meant that an increase in African doctors 'to help their own people' would not solve the problem, because the issue was not one of race and language, but of belief and culture. The issue arose because *'the mass of the Natives in the Reserves* ... still greatly prefer diviners and herbalists to scientifically trained doctors', whether white or black. 'One would', he continued, 'be sorry to see a third medical school differentiated from the other two on the ground of the colour of its students'. Instead, he advocated a 'special school' for 'Native Health Officers' in Durban, whose role would be as much health education as disease prevention, together with increased government support for overseas bursaries for Africans who wished to study overseas.[46]

Once in government service in Pretoria and perhaps more in touch with educated African opinion on the Rand, Gale's ideas moved somewhat. No longer overtly rejecting the segregationist practices of his day, by 1943 he justified the creation of a full 'non-European medical school' in Durban, 'by reason of its situation amidst a diverse population and an industrial revolution'.[47] Gale's words may have been chosen to chime in with those of his correspondent, the Principal of Natal University, but his commitment to Durban was already recorded before the Botha Commission on Medical Education in 1937. Moreover, he believed that in Durban students would encounter in both rural and urban settings the diseases which most afflicted Africans – malaria, bilharzia, tuberculosis and the dysenteries.[48] At the same time the political implications of establishing a medical school and choosing Durban (where there was no white medical school) were also relevant:

> The segregation principle which dominates the socio-economic pattern of South African life demands or at least implies, the provision of liberal opportunities for the educated Bantu to serve their own people. To do this on a big scale – a medical and health service staffed largely with

Native personnel – would do much to restore the faith of the Bantu in their European rulers.[49]

This pragmatic preparedness to work within the conventional segegrationist wisdom of the time may also have made Gale more acceptable to the Department of Health. In the late 1930s Cluver was only too aware that proposals for black welfare could – as he put it – 'be shipwrecked by malevolence or even over-enthusiasm for Native cultural development'.[50]

Yet Gale's 1938 pamphlet can be read as a more unequivocally progressive document in its recognition of the social basis of ill health and its advocacy of health education and preventive medicine in the rural areas. It was this that resonated with the very different tradition out of which Sidney Kark was emerging in the 1930s at the University of Witwatersrand's Medical School, where at a young age he initiated a 'Society for the Study of Medical Conditions among the Bantu' and where he was a leading member of the students' Socialist Party. It was there that Henry Gluckman lectured, for example, before going into parliament as the only way to further his concern with health and welfare; it was there that such pioneers of nutritional studies as Theodore and Joe Gillman were at work, and where Kark and David Landau, another pioneer of social medicine and the health centres, met in the left wing of the Labour Party.[51] The number of Jews involved in progressive medicine at this time is notable, a product of the disproportionate number of South African Jews who practised medicine in the 1930s and the equally disproportionate – if still small – number of Jewish political radicals.[52]

As in other fields, medical discourse at the university was influenced by international developments in social medicine in the UK and USA. In this context, the visit in 1937–38 of Henry Sigerist, described in the *Cape Argus* as the 'foremost apostle of socialised medicine',[53] as a guest of the South African Students' Visiting Lecturers' Trust Fund, further stimulated and inspired student radicalism.[54] His lecture tour was but the most visible tip of a considerable shift to social medicine in the 1930s and 1940s. A later more disillusioned generation may read into the social medicine of these years an increase in the state's 'disciplinary' powers and its capacity to screen and 'surveil' its subjects through increasingly subtle modes of control. Yet this is at best only a partial truth. Social medicine also had within it a liberatory potential which inspired and excited the younger, more radical members of the medical community, surrounded as they were by the disease consequences of rapid industrialisation and rural impoverishment.[55]

This is not to say that the 'conditions of possibility' for such a use of coercive power did not exist. Health centres were undoubtedly designed to intervene in family and home life and if need be to remodel it, through the agency of specially trained 'native health assistants', whose role was seen as primarily health education. A major part of the training for health assistants in assessing a community's 'needs' was concerned with 'finding out what the community feels, thinks and does about its health needs' through surveys, group discussions and home visits.[56] And while this seems in tune with contemporary notions of community participation, and the advice they offered purely technical – concerning clean water, sanitation and nutrition – it can also be seen as highly prescriptive. 'We must', Gale declared in 1937, 'have *propagandists* to meet the dead weight of ignorance and superstition, to carry the battle right into the very homes of the people.'[57]

Nevertheless, as Dorothy Porter has argued in relation to Britain, social medicine never acquired the hegemony the Foucauldian model presupposes;[58] indeed, in most respects the evidence suggests that doctors in the forefront of social medicine were also in the forefront of advocating a more liberal and liberating medical practice. For example, when in the mid-1930s the government appointed Dr Harry Gear as Union Assistant Health Officer to develop programmes on venereal diseases, he argued that combining health examinations with measures to control African urban settlement

> went against a basic principle that a health programme should be based on the co-operation and trust of the public. Africans treated for VD would conclude that, especially in its association with police, pass issues and other official relationships ... medical and health measures are not for the benefit of the natives, but for repressive and disciplinary purposes.[59]

Cluver, at that time Secretary for Public Health, also maintained that health schemes which might 'antagonise or ... intrude "police methods" was of "extremely doubtful utility"'.[60] In South Africa there were always far more direct and coercive ways of controlling colonial bodies and conducting surveillance in the practices of mine medicine and mental hygiene than among the acolytes of social medicine.

The National Health Services Commission

The war years, which saw such rapid urbanisation and industrialisation and a concomitant increase in the burden of ill health, also saw a further and more general radicalisation of the medical profession in

South Africa, as in so many other spheres of social and professional life.[61] The Federal Council of the Medical Association of South Africa (MASA) set up a planning Committee to consider a future national health policy for the Union, the *South African Medical Journal* carried a series of articles on the subject, MASA published a pamphlet advocating a national health service controlled by the medical practitioners and, partly in response, in 1942 the government set up a 'National Health Services Commission on the Provision of an Organized National Health Service for all sections of the People of the Union of South Africa'.[62] The very title is noteworthy. Its recommendations involved the total reorganisation of South Africa's health services.

As a civil servant, Gale was not a member of the Commission, but, as Gluckman's confidant and adviser, he greatly influenced its outcome. One should also not underestimate Gluckman's own role, however, and what Gale called 'a very happy conjunction of circumstances that there were on the Commission men very receptive to progressive ideas, and at the same time spokesmen for the organised profession [MASA] who furnished them with progressive ideas and powerful arguments in support of them'. Men like Cluver and Harvey Pirie were chief spokesmen for MASA, members of its Planning Committee and largely responsible for the memorandum it presented to the Commission.[63]

Nevertheless, it was through Gale that Kark and Gluckman met, and Gale brought Kark's work at Pholela to the attention of Gluckman and the Commission.[64] On learning about Pholela, Gluckman is said to have remarked: 'This is the model for the native territories.' By this time, however, Gale, who had started out with a concern for the health needs of rural Africans saw the health centre not simply as a model for the 'native territories, but also for the whole country'. He was aware, as Yach and Tollman remark, that 'change in the health care system cannot develop out of a program solely limited to the poor', because a project for the poor and powerless would itself be poor and powerless.[65] At the same time he believed the model being developed at the health centres offered 'a great opportunity to develop social medicine for masses of people who are rapidly changing their economic and cultural patterns' and to learn lessons 'of interest and value to other countries where similar changes are taking place'.[66]

Between 1946 and 1948 nearly forty health centres were established in South Africa, generally in the poorest rural and urban areas: notably this included poor white and coloured as well as African areas. Each health centre served a defined area within which staff conducted home visits; it also provided a casualty service for people living beyond who came of their own volition. In addition, centre staff helped local people with 'simple environmental sanitation' and stimulated the establishment of school feeding schemes, nursery schools, recreation clubs, gardening clubs and discussion groups. Crucial to the health centre scheme was the definition of a circumscribed area within which reliable statistics could be collected so that improvements in mortality, morbidity and living conditions could be measured, and the outcome assessed of a programme of health education carried out over a number of years. An epidemiological approach was essential in the absence of any national registration of black births and deaths. In addition, through the keeping of accurate records, this strategy could also be shown to be cost effective – 'despite criticism of it as the luxurious fad of mere doctrinaire'.[67]

Ultimately the Commission planned on the establishment of some 400 centres, each servicing a population of 25,000, regardless of race. Each centre was to have six to eight doctors with nurses and other auxiliaries. These were not an end in themselves. They were intended to form part of a far wider vision. As the foundation of a truly comprehensive national health service, and 'the first step in the implementation of... [its] recommendations... which also involved the nationalisation of all personal health services,' they would distribute doctors and other health personnel more rationally and efficiently.[68] Unfortunately, however, as personal health services were not nationalised when the health centres were established, so the health centres, intended as Gale remarked as 'only a first step in the establishment of a national health service... proved also to be the last'.[69]

The conventional wisdom, then and now, is that this was simply the result of the National Party victory in the 1948 elections. Thus, according to Gale, the health centres were 'ultimately undermined by political action'. 'Having been introduced by one Government, they were looked at askance by the government which displaced it.'[70] While the hostility of the National Party to the health centres was considerable, however, this was not the only, nor even perhaps the main, reason for their failure, as I have tried to show elsewhere.[71]

As early as 1944, even before the NHSC Report had been published, the Prime Minister, Field Marshal Jan Smuts, had given way to the clamour of the politically powerful provincial authorities (which were then responsible for running curative services) and allowed hospital services to remain in provincial hands.[72] Gale himself realised at the time that even Smuts's government 'did not wholeheartedly accept, or at least act upon, the findings of its own Commission'. After the war, hospital expenditure escalated from ten million to forty-five million dollars a year, while the government provided under a million dollars 'to start some Health Centres in an experimental sort of way. The attitude of the present [i.e. Nationalist] Government', he added, 'is much about the same. It is very difficult for any government to resist the urgent plea for more hospitals to meet immediate urgent needs, and no less difficult for it to invest funds in services which do not give immediately obvious returns.'[73]

Nor did the Smuts government accept the major recommendation of the Commission that the national health service be financed through a direct health tax: given the unpopularity with the electorate of increased taxes, this was too radical a recommendation for a somewhat beleaguered wartime government. Personal curative care remained in the hands of private practitioners except – as in the past – for the indigent. The improved health of whites as a result of the economic boom of the early 1950s and National Party affirmative action for its Afrikaner supporters meant that white ill health receded from their political agenda. Their poorer constituents, now largely cared for by medical benefit schemes, no longer needed cheaper health care, while the growing number of Afrikaner doctors (like their English-speaking counterparts) were positively hostile to any encroachment on their private practice. This meant that, contrary to the original intention, the health centre idea was indeed increasingly confined to the poor and the black.

The wartime idealism of the medical profession which had depended on the vision of a small number of leading men, had proved remarkably short-lived, while a number of teething problems, including the absence of trained professionals to run the new health centres, gave opponents of the health centres their ammunition.[74] As early as February 1948 Gale found it necessary to counter criticism that there was no 'colour bar' in the training scheme for centre personnel.[75] In a visit to South Africa as a guest of the South African National War Memorial Health Foundation in 1948, the eminent first Professor of Social Medicine in Oxford, John Ryle, whose ideas in part informed the NHSC Report,[76] also voiced his criticisms of the health centres despite his sympathy for their overall objectives. However well-intentioned, his criticisms – which reflected his concern to link social with clinical medicine and ensure its academic respectability – probably exacerbated, rather than resolved, the problems facing the health centres.[77]

It is notable that where the health centres were functioning satisfactorily it was possible for them to be saved even in the uncongenial environment of South Africa after 1948. Thus, in 1954, the Department of Health was still able to protect those health centres which were run efficiently and economically, and in the following year

Norman Reeler (who had worked in the department with Gluckman and Gale, and had been the Secretary of the NHSC) was able to save the Lady Selborne Health Centre in Pretoria's African township from abolition on these grounds.[78] It is clear, nevertheless, that Nationalist ministers had little enthusiasm for 'social medicine' and the health centres and the Institute for Family and Community Health were under constant scrutiny. As Gale was to write to his friends in 1955, 'From 1948 onwards I had repeatedly to exercise all my official powers as Secretary of Health and all my personal powers of persuasion with successive Ministers, in order to ensure the continuance of the Institute: even so, I could not save it from considerable mutilation.'[79] In 1950, opponents of the scheme within the department took advantage of Gale's absence overseas to persuade the Minister of Health to appoint a wholly hostile 'committee of inquiry' into the health centres, which proposed their abolition. Although Gale was able to fend this off on his return, he realised that this effectively spelt the end of the health centres and that he was merely fighting 'a rearguard action'.[80] By 1953 many of the remaining thirty health centres were being integrated into the district surgeoncies.[81]

Gale and the Durban Medical School

By this time, however, George Gale had left the Department of Health to become Dean of the newly founded Natal Medical School, with supplementary funding from the Rockefeller Foundation. Long interested in the medical education for Africans, he had already played a major role in the School's establishment in Durban against the rival claims of Johannesburg. To ensure this outcome, he had intervened directly to persuade Gluckman of the merits of a Durban base and together they had convinced Smuts of their case, against the wishes of the minister responsible, Jan Hofmeyr, in the last days of the United Party government.[82] More importantly he also managed to dissuade the first Nationalist Minister of Health, Dr A. J. Stals, from following the wishes of the government to relocate the new medical school in the heart of the Transkei or Ciskei, because of the allegedly 'baleful' influence of the city on the 'Bantu'.[83]

Although Gale's 1938 pamphlet had accepted that white racism would prevent the medical training of black South Africans, by the late 1940s the Universities of Cape Town and the Witwatersrand were training very small numbers of black doctors. Nevertheless, Gale had a very different vision of appropriate medical training in mind: as the Principal of Natal University, E. G. Malherbe, wrote in urging him to take the post: 'I have all along felt that you would be able to give the medical school a unique orientation and would thus make a unique contribution to health and medical training in South Africa, and possibly the whole of Southern Africa.'[84] Under Gale, the new Medical School would not simply be 'a rather feeble replica of what already exists in the other medical schools'. As Malherbe understood, to address the needs of health in Africa, medical education had to be directed away from a purely curative, hospital-based approach towards what Gale called 'social medicine, the medicine of the future'.[85]

This was a cause close to Gale's heart. As early as 1947 he and Kark had realised that by the time young doctors arrived for training in social medicine it was too late: according to Gale social medicine was 'an attitude rather than a subject, and should be inculcated during the undergraduate phase'. At that time he had tried to persuade the Rockefeller Foundation to subsidise the young Professor Brock, who had been Ryle's first assistant in Cambridge, in setting up a social medicine unit at the University of Cape Town. This does not seem to have succeeded, in part perhaps because the Rockefeller advisers feared that supporting social medicine in Cape Town would not have 'an important bearing on the greatest of South African social problems since the Negro population is not as significant in this area as it is elsewhere'.[86]

This accusation could not be levelled at Durban, where Gale hoped to establish a training school 'in Africa for African conditions': he therefore wished entry to be open to *all* South Africans as well as to students from other parts of Africa. Aware of the strong antagonism to such a course on the part of the National Party government, he was prepared to accept that white undergraduates applying to study at Durban would have to be sanctioned by the Minister of Education but hoped that postgraduates would have no such difficulty. Despite this compromise, however, he was soon locked in conflict with the government over admissions policy.[87]

Nor was this the only source of conflict. Even before Gale's deanship, there had been constant tussles between the University and the Treasury over money, as the original estimates for building and equipping the Medical School proved inadequate. By the end of 1953 the Principal of Natal University, E. G. Malherbe, was apologising for a 'misunderstanding by the Dean [Gale] of the proper procedure' which placed the Treasury 'in an embarrassing situation': the Dean had spent considerable sums on the Medical School without prior approval. Clearly Gale had a very different conception of an adequate Medical School to that of his political paymasters: he was adamant that medical training for black students should be the very best and most progressive available. To 'secure the confidence of the non-Europeans', he believed it was 'vital' to create a School 'in no way inferior . . . [to] the other schools in the Union': 'To achieve this at a time when the Government was making lawful the provision of separate and inferior facilities for non-Europeans, was my major objective throughout my three-year tenure of the Deanship.'[88]

In 1954 a two-man committee was appointed to look into the finances of the University and the Medical School. It recommended savings in the expenditure on both staff and equipment, and dismissed the need for a full-time Dean – on the grounds that the much larger faculties in Pretoria and the Witwatersrand had no such full-time appointee.[89] Undoubtedly part of the animus was against Gale personally and against the intervention of outsiders like the Rockefeller Foundation.[90]

The end of a dream

Given his financial insecurity, by January 1955 Gale had accepted an invitation to become the first Professor of Preventive Medicine at Makerere University College in Uganda. As he saw it:

> The Wilcocks-du Toit Report [on the funding of the Medical School] makes clear that a policy of ruthless and unimaginative economy is henceforth to dominate the relations between the State and the Durban Medical School and, therefore, the 'somewhat extravagant luxury' which I am described as being has no choice but to accept an alternative post as soon as it offers, rather than remain – an embarrassment to the University vis-à-vis the Government – until such time (1957 at the latest) as the government sees fit to pronounce sentence of death.[91]

Gale's subsequent career does not concern us here. Very briefly, while at Makerere, he established a 'Pholela-type' health centre at Kasangati, and was active on Uganda government commissions and in work for the WHO. After his retirement in 1960, he spent four years in Thailand and another four in Malaysia as Visiting Professor of Social and Preventive Medicine, supported by the WHO. In Thailand he drew up curricula in social medicine for the two medical schools in Bangkok, and the new medical school in Chieng Mai. It is perhaps no coincidence that Thailand and Malaysia today are reputed to have major primary health care programmes, some of which bear an uncanny resemblance to Pholela.[92]

Gale's departure marked a watershed in the short history of social medicine in South Africa. For Sidney Kark, 'It was the end of a dream.'[93] The Dean's departure made the entire social medicine project more vulnerable. By 1957 the Karks also felt impelled to leave, fearing a takeover of the medical school by the state under the impending legislation would intensify the harassment they and their colleagues had experienced while they were part of the Union Health Department.[94] After a year in North Carolina, Sidney Kark took up a WHO-funded professorship in Jerusalem where he was joined by many of his Clairwood team.

In the 1940s both Gale and Kark were able to make use of a rare window of opportunity to initiate and develop far-sighted policies of social medicine. That 'window of opportunity' was in part the result of the war and the brief period of reformism that resulted from it – the reflection in South Africa of the 'welfare corporatism' that underpinned reconstruction after the war in Europe.[95] It was also in part the result of a crisis in black health – which it seemed could be resolved more cheaply for white taxpayers and provincial authorities through health centres than hospitals – which coincided with a crisis in white health. Momentarily this coincidence lent the health centre idea wider support through the Gluckman Report. Once the urgent need for reform for whites disappeared, however, and the balance of forces in the state shifted, the impetus behind social medicine coming from within the Department of Health diminished.

Enemies of the scheme were able to use weaknesses in its implementation to undermine the health centre movement, while after the war the conservatism of the medical profession quickly reasserted itself and further subverted it. At the same time the bitter anti-communism of the Nationalists, strengthened as it was by the Cold War, intensified the hostility to notions of 'social medicine' with its connotations of 'socialised medicine'. With the dispersal of many if not most of the doctors associated with the health centres, the social medicine movement was killed in South Africa – even if as an unforeseen consequence its ideas spread to Israel, the United States and Asia. They were not to be revived in South Africa until the reformism of the 1980s and the advent of a post-apartheid government in 1994.

In retrospect, it is perhaps not surprising that Gluckman's recommendations of a health service based on the social medicine ideas of the health centres were dropped after 1948. The central dictum of the Commission – that unless there were drastic reforms in the sphere of wages, nutrition, housing, health education and recreation, 'the mere provision of more "doctoring"' would 'never bring health to the people of South Africa' – demanded a drastic restructuring of the social order, which went well beyond the white consensus.[96] Perhaps it also exceeded the capacity of the South African economy, which was still so heavily dependent on the low-wage sectors of farming and mining. Dominant whites in the state were simply not prepared to sustain the welfare costs involved in a National Health Service. The Afrikaners who gained power in 1948 were even more dependent on exploiting cheap migrant labour. Once the immediate problems of poor-white health were resolved and improved, and expanded benefit societies and insurance schemes (a crucial area of Afrikaner accumulation) were able to take care of white working- and middle-class health, the impetus for a more radical solution of South Africa's health problems evaporated, until the political configuration began to change in the 1990s.[97]

Notes

1 I am most grateful to Dr George Gale's son, Mr Roy Gale, for his detailed comments on an earlier version of this chapter, and for allowing me temporary possession of the Gale papers pending their transfer to an appropriate library. I am also grateful to the Rockefeller Foundation for a grant enabling me to consult their archives in Tarrytown, NY, and for the kind assistance of their highly skilled archivists.
I have used the following abbreviations in the footnotes: *AJPH* American Journal of Public Health; GES Department of Health, Pretoria; *JSAS* Journal of Southern African Studies; MASA Medical Association of South Africa; NHSC National Health Services Commission; NTS Department of Native Affairs, Pretoria; PH Public Health; RF Rockefeller Foundation (Archives); SAB South African Archives Bureau (Pretoria); SAIRR South African Institute of Race Relations; *SAMJ* South African Medical Journal; Sec. Secretary; SNA Secretary for Native Affairs.
2 Gale Papers: (Mrs) Audrey I. Gale, 'Outline notes on George William Gale', 4 February 1981. It is possible that the health centre idea was brought to South Africa by Dr Harry Gear who had encountered Grant's health centres as a young doctor working in Shanghai. Gear was behind the establishment of the first South African health centre at Pholela (Author's interview, Sidney and Emily Kark, Jerusalem, 29 September 1982, henceforth Kark interview).
3 The quotation is in P. V. Tobias, 'Henry Gluckman', *SAMJ*, 72 (15 August 1987), 303.
4 D. Porter, 'Social medicine and the new society: medicine and scientific humanism in mid-twentieth century Britain', *Jnl of Historical Sociology*, 9:2 (1996).
5 University of Natal Archives, Pietermaritzburg, and Gale Papers, 'The story of the Durban Medical School – told by G. W. Gale', TS, 25 January 1976. This account was written shortly before Gale's death to assist in the defence of Durban Medical School then threatened with abolition. For the Karks' engaging account of their experiences at Pholela and the Institute of Family and Community Health, see S. and E. Kark, *Promoting Community Health: From Pholela to Jerusalem* (Braamfontein, 1999).
6 See, for example, Quentin D. Young, 'Health care reform: a new public health movement', and H. J. Geiger, 'Community oriented primary care: The legacy of Sidney Kark', editorials in *AJPH*, 83:7 (1993), 945–7; T. H. Phillips, 'The 1945 Gluckman Report and the establishment of South Africa's health centers', M. Susser, 'A South African odyssey in community health: A memoir of the impact of the teachings of Sidney Kark', and D. Yach and S. M. Tollman, 'Public health initiatives in South Africa in the 1940s and 1950s: Lessons for a post-apartheid era', all in the section entitled 'Public health then and now' in *ibid.*, 1037–50; M. Susser, 'What is community oriented family care?' unpublished talk at the Jerusalem Memorial for Sidney Kark, January 1999.
7 See Kark and Kark, *Promoting Community Health*, pp. 10, 178–9.

8 This was the view of his wife and family, and of the Karks, although I have found no written confirmation (Kark interview; author's interview with Audrey Gale, 21 June 1983; pers. comm. to author from Roy Gale, 3 August 1994).
9 Ibid.
10 Emily Kark to Audrey Gale, 7 March 1978, kindly shown me by Mrs Gale, 1982.
11 RF RG 1.2 Series 487A Box 3 Folder 25, Gale, CT, to Dr George C. Payne, NY, 11 April 1950.
12 See Gale, 'Story of Durban Medical School', pp. 13–14; Gale Papers, Gale to E. G. Malherbe, 20 August 1951 (applying for the position of Dean); and RF RG 1.2 Series 487A Box 2 Folder 19 Item E5544 WAH Resolution of the Executive, 24 September 1954, approved 26 July 1955. In the event, the funding took effect after Gale left in 1956 because the South African government dragged its feet (ibid).
13 See Vanessa Noble, 'Medicine and power: a history of the University of Natal Durban Medical School', unpublished paper presented to the African history seminar, University of Natal, 24 November 1998.
14 Gale-Aitken Correspondence, York University microfilm, Gale to R. D. Aitken, 21 September 1970. I am grateful to Lisa Gale (no relation) for finding and copying this letter for me.
15 See George Gale, *A Suggested Approach to the Health Needs of the Native Rural Areas of South Africa* (privately printed by the author, Benoni, 1938), p. 15.
16 See Gale Papers, 'The medical education of non-Europeans in South Africa', n.d. c. 1968. For Fort Hare, see below.
17 Gale, 'The medical education', p. 24.
18 His name, for example, is together with those of a number of prominent liberals on a flier in October 1952 calling on the government to take heed of the Defiance Campaign and extend equal rights to all 'civilised men'.
19 Pers. comm. Roy Gale 3 August 1994.
20 For a critique of South African doctors and their propensity to adopt these positions, see Randall Packard, *White Plague. Black Labor: Tuberculosis and the political economy of health and disease in South Africa* (Berkeley, 1989), pp. 235–47.
21 'Falling off in Native Health. Evidence before the Commission', *Star*, 11 October 1943. See also his evidence before the UG 28–'48, *Native Laws Commission* (the Fagan Commission) (Pretoria, 1948), pp. 35–9.
22 Cullen Library, Wits. AD 843 B8.1.6 HPW, Dr G. W. Gale, 'Health services in the Union' (n.d. c. 1946–47), cited in K. Jochelson, 'The colour of disease: Syphilis and racism in South Africa, 1910–50', forthcoming, p. 262.
23 Pers. comm. Roy Gale, 3 August 1994.
24 See RF RG2 1948 Series 487 Box 423 Folder 2857. GKS, [?Struthers] Diary Excerpt, 18 October 1949. The observer was Dr D. B. Wilson who visited South Africa with Grant in September 1948, representing the Foundation.
25 George Gale, 'Government health centres in the Union of South Africa', *SAMJ*, 23 (23 July 1949), 632.
26 Charles Rosenberg, *Explaining Epidemics and other Studies in the History of Medicine* (Cambridge, 1992), p. 260.
27 See *NHSC Report*, p. 41, para. 25. See also Gale, *Suggested Approach*, and George Gale, 'Native medical ideas and practices in relation to native medical services', *SAMJ* (27 October 1934), and 'The rural hospital as an agent in native health education', *SAMJ* (8 August 1936), 541–3.
28 For the Karks' attempts to establish self-sufficient community organisations, see Sidney and Emily [Kark] to Mervyn [Susser], 20 August 1987. I am grateful to Professor Susser for sharing this personal letter with me.
29 George Gale, *John William Bews* (Pietermaritzburg, 1954), pp. 93–4. As Dorothy Porter shows in her 'John Ryle: Doctor of revolution?' in D. and R. Porter (eds), *Doctors, Politics and Society: Historical essays* (Amsterdam, 1993), pp. 258–61, Smuts's philosophy of holism was adopted by a number of researchers working in the 'biotheoretical sciences' in the 1930s. See also her 'Social medicine and the new

society', pp. 175–6, for the ways in which Smuts's ideas influenced John Ryle, founder of social medicine in Britain.
30 Audrey Gale, 'Outline notes'; CV attached to Gale to Malherbe, 29 August 1951.
31 Letter from Roy Gale, Otford, Kent to S. M., 20 June 1994: 'He undoubtedly regarded all his medical work including the eleven years' study after schooling as an application of his very strong conviction as a Christian from a very early age.' Also telephone conversation, 29 June 1994.
32 Gale Papers *passim*, especially his MS address to the Students' Christian Association on 1 November 1919, his letters to his family in South Africa from Scotland in 1924, his letter to Dr Frazer from the Gordon Memorial Mission, 2 March 1931, and the letter from the Bishop of Uganda, 15 March 1960.
33 Historical Papers, Wits. Hofmeyr Papers, A1 Aa 1937, Reg Pearce to 'Hoffie', 27 February 1937.
34 Gale detailed his criticisms in his *Suggested Approach*, pp. 15–25, and Gale, 'The medical education'; see also Karin A. Shapiro, 'Doctors or medical aids – the debate over the training of black medical personnel for the rural black population in South Africa in the 1920s and 1930s', *JSAS*, 13:2 (1987).
35 Pers. comm. Roy Gale, 5 August 1994.
36 Gale, 'The medical education'.
37 Gale, 'Story of Durban Medical School'. According to Gale, 'The medical education,' p. 10, he was the one who persuaded Cluver and Gear that the first health centre should be located at Pholela in Natal because he argued it could 'in due course make an important contribution to a medical course to be developed at the non-European medical school which the Botha Committee [1938] had recommended be established in Durban'.
38 SAB, A 1207, B2.2 Confidential memorandum by G. W. Gale, 'Health Centres Service of the Ministry of Health', circulated to members of the National Health Council and Advisory Committee on Health Centre Practice, February 1948.
39 Kark interview; Audrey Gale, 'Outline notes'.
40 According to Roy Gale, his father was converted to socialism while a student in Scotland in the 1920s. Author's interview with Roy Gale, 1 November 1995.
41 See Gale, *Suggested Approach*, pp. 26–9.
42 See GES 155 1/62, 'Native Health and Hospital Services. Notes of a sub-committee of Provincial Consultative Committee held in Pretoria on 7th November, 1938'. The quotation is in an enclosed letter dated 4 February 1939.
43 See, for example, Gale, *Suggested Approach*, pp. 5–8.
44 *Ibid.*, p. 7.
45 *Ibid.*, p. 8.
46 *Ibid.*, pp. 2, 11, 13 (Gale's emphasis). He feared that the lack of adequate funding for such a medical school would result in 'an inferior medical qualification on the grounds of colour'.
47 Gale to Malherbe, 29 August 1951.
48 Killie Campbell Library, Mabel Palmer Papers, KCM 17422, File 20, 'Copy of a memo received from Dr Gale – submitted to the Public Health Commission, 1943'.
49 Gale Memo on Durban Medical School, 1943.
50 SAIRR B 91.2.1 Health Papers: Cluver, Department of Health, to A. Hoernlé, 9 June 1934; Cluver was writing about the plans for training medical aids, but the point remained equally valid five years later.
51 A friend of Gale's, Landau became head of the Social Medicine Division of the Department of Public Health in Natal and was deeply interested in Pholela. According to Kark, his personality was one of 'brilliant excitement'. Tragically he died in August 1948, at the early age of 42. (Kark interview.)
52 For the ambience at Wits Medical School in the 1930s, see the note from Sidney and Emily Kark to Mervyn Susser, 20 August 1987 (see n. 28 above). For the number of Jewish doctors, see the [Botha] *Report on Medical Training in South Africa* (Pretoria, 1939); and for Jewish radicalism, I. Suttner, *Cutting through the Mountain: Interviews with South African Jewish activists* (London, 1997).

53 'No future without state medicine... World scientist gives his conclusions', *Cape Argus*, 15 December 1939. For Sigerist, see Elizabeth Fee and Edward T. Morman, 'Doing history, making revolution: The aspirations of Henry E. Sigerist and George Rosen', in Porter and Porter, *Doctors, Politics and Society*, pp. 275–311.
54 Gale and Kark agreed on the importance of Sigerist's visit. (See Gale Papers, Gale to 'My dear Eustace [Cluver]', 8 October 1967 and Kark to Gale, Airletter Card, no address or date, c. November 1967.) For social medicine in Britain at this time, see Porter, 'Social medicine and the new society'.
55 The increase in community surveillance through social medicine and its ancillary, the social survey, is the central burden of David Armstrong's critique in *The Political Anatomy of the Body: Medical knowledge in Britain in the twentieth century* (Cambridge, 1983), esp. pp. 32–41. For its application to social medicine in South Africa, see Alexander Butchart, *The Anatomy of Power: European constructions of the African body* (London and New York, 1998), pp. 138–50. For a counter view, see Rosenberg, *Explaining Epidemics*, p. 260.
56 GES 2745 65/70, G. W. Gale, 'The training of health assistants as health educators', n.d. November 1951.
57 Gale Papers, Gale (Municipal Health Dept, Pietermaritzburg) to Dr Fox, 28 November 1937.
58 Porter, 'John Ryle', pp. 263–5.
59 GES 376 4/5B Sec. for PH to SNA, 23 January 1939, cited in Jochelson, 'The colour of disease', p. 262.
60 NTS 5761 15/315 vol. 3, in *ibid*.
61 For the rising tide of ill health, see S. Marks and N. Andersson, 'Industrialization, rural health, and the 1944 National Health Services Commission', in Steven Feierman and John M. Janzen, *The Social Basis of Health and Healing in Africa* (Berkeley, Los Angeles and London, 1992).
62 UG 30 – 1944. See also Marks and Andersson, 'Industrialization, rural health', pp. 154–5.
63 Gale Papers, Gale to 'My dear Eustace [Cluver]', 8 October 1967.
64 Kark interview; Gale Papers, TS Copy of 'Aerogram II of Sept 20 1967 from GWG': 'Whether Gluckman visited Polela or not, he read the reports (and although I shall not say this! listened to G.W.G.)...' and Gale to 'My dear —' cyclostyled letter, 'Personal and confidential', from Kampala, 21 May 1955.
65 Yach and Tollman, 'Public health initiatives', p. 1047.
66 RF RG Series 487 Box 385 Folder 2600, Gale to Grant, 27 September 1947.
67 'Health Centres in South Africa: An obituary', Part of MS found in the Gluckman Papers, Wits. (henceforth 'An obituary'). This statement may have been directed at Ryle for Gale continued: 'Too often health centres have been judged on the basis of "impression data" – the impressions formed by visitors who are inevitably biassed by their own preconceptions regarding "social medicine" and... may also be misled, consciously or unconsciously, by those who show them around.'
68 'An obituary'.
69 *Ibid*.
70 *Ibid*.
71 See S. Marks, 'South Africa's early experiment in social medicine: its pioneers and politics', in 'Public health then and now', *AJPH*, 87:3 (1997).
72 The powers of the provinces had been safeguarded by South Africa's Act of Union but his concession may have reflected the weakness of a wartime government controlling a divided population, rather than the absolute strength of the provinces in relation to central government.
73 RF RG 1.2 Series 487A Box 3 Folder 25, Gale, CT, to Dr George C. Payne, NY, 11 April 1950.
74 The similarity to the fate of post-1994 health policies is striking.
75 SAB A 1207 B2.2, Confidential memo, 'The Health Centres Service of the Ministry of Health', p. 21.

76 According to Gluckman, who was speaking at the Natal Technical College, 2 March 1948, in the presence of Ryle who was the guest speaker. H. Gluckman, *Abiding Values: Speeches and addresses* (Johannesburg, 1970), pp. 108–9.
77 For Ryle, see Jane Lewis, *What Price Community Medicine* (Brighton, 1986), especially pp. 37–8, and Porter, 'John Ryle'. I have not traced Ryle's report to Gale, but there are two letters in the Rockefeller Archives to John Grant and Gregg which convey its gist (RF RG2-1948 487, Box 423 Folder 2857, Ryle to Grant, 17 June 1948 enclosing Ryle to Gregg, 31 March 1948): See also Gale, 'Story of Durban Medical School', pp. 7–8.
78 GES 2094, J. J. du P. le Roux, Secretary of Health to Minister (A. J. R. van Rhijn), 29 September 1954; and GES 2734 50/23/70, Reeler to Sec. Public Service Commission, 25 January 1955.
79 Gale to 'My dear —', 21 May 1955.
80 Gale, 'The medical education', p. 23.
81 WHO/PHA/21 Expt Cttee on PHA, B. M. Clark, 'Health services and health centres in the Union of SA'.
82 Gale, 'The medical education'; M. E. Beale, 'Apartheid and university education, 1949–58', MA thesis, University of the Witwatersrand, 1994; Vanessa Noble, 'Medicine and power: A history of the University of Natal Durban Medical School', unpublished paper, University of Natal, African Studies Seminar, 1998; see also Killie Campbell Library, KCM 56990 (38) Malherbe Papers, A. Gale to E. Malherbe, 3 April 1976.
83 Gale, 'Story of Durban Medical School', pp. 8–10.
84 Gale Papers, Malherbe to Gale, 29 September 1951.
85 *Ibid*; and Gale to Malherbe, 29 August 1951.
86 RF RG2 1947 Series 487 Box 385 Folder 2600, Gale, Pretoria, to Grant, NY, 27 September 1947; RF RG2-1948 Series 487 Box 23 Folder 2857, Brock to Davie, 9 October 1948, and 'Memorandum on the teaching of social or preventive medicine in the Medical School of the University of Cape Town'. *Ibid.*, Box 423 Folder 2857 Diary R. S. M[orison?] Interview, 28 October 1948: J. B. G[regg].
87 Gale, 'The medical education', p. 29.
88 Gale to 'My dear —', 21 May 1955.
89 SAB TES 2097 8/238, Malherbe to Sec. for Arts, Education and Science, 29 December 1953; *Ibid.*, U. 3/40/4 'Report on the Medical Faculty of the University of Natal' (the Wilcocks-du Toit report).
90 See RF RG2 1948 Series 487 Box 423 Folder 2857, Diary of a trip, D. B. W[ilson], p. 2, 25 September 1948; Gale, 'The medical education', pp. 28–9.
91 TES 2097 8/238, Malherbe to Sec. for Arts, Education and Science, 12 January 1955.
92 This somewhat speculative generalisation is based on my own limited observation of a rural primary health care centre in Udom Thani. Gale's impact in South Asia – he was the inspiration of some health centre work in India and acted as external examiner in Hong Kong – awaits further exploration.
93 Kark interview.
94 RF RG 1.2 Series 487A Box 2 Folder 20, Kark to Malherbe, 4 April 1957. Kark complained that during that time they had been 'subjected to constant investigations and criticisms, and ... faced recommendations by various commissions that were aimed at our abolition'.
95 Porter, 'Social medicine and the new society', pp. 172–3.
96 See chapter XX of the Gluckman Report; the quotation is on p. 133. According to the Report, this view was emphasised by many witnesses, 'not least by the official representatives of the Medical Association of South Africa'.
97 See Marks and Andersson, 'Industrialization, rural health', for a further elaboration of this argument.

CHAPTER NINE

Technical development and the human factor: sciences of development in Rhodesia's Native Affairs Department[1]

Jocelyn Alexander

Throughout British colonial Africa, the late 1940s inaugurated what D. A. Low and John Lonsdale have memorably termed the 'second colonial occupation'.[2] This was a period in which great faith was placed in the merits of state planning, in the possibility of increasing economic efficiency through technical innovation, and in close government regulation. The expansion of state activity relied on the elaboration of new scientific practices and expertise, and in the institutionalisation of cadres of experts. It heralded an unprecedented intervention into the ways in which Africans lived and farmed. In Southern Rhodesia, this was the era of 'technical development'. The bright promise of technical development did not, however, last long. Its rejection in favour of what was dubbed the 'human factor' in the early 1960s marked a period of dramatic transition in Rhodesia's sciences of development. A study of this transition allows an exploration of a number of important themes in the study of colonial science: the divisions and debates within colonial bureaucracies; the relationship between scientific knowledge, practices and power; and the interaction between scientific authority and political crisis.

I begin with a brief introduction to the elaboration of technical development policies in Rhodesia before turning to the rise and demise of the most elaborate version of this trend in developmental science, the Native Land Husbandry Act of 1951, and close with the emergence of a new science of development and control in the 1960s.

Technical development

The institution charged with administering and developing Rhodesia's African reserves was the Native Affairs Department (NAD). The NAD enjoyed wide-ranging legislative powers and autonomy with regard to

[212]

Africans and acted as the self-proclaimed guardian of African interests within government. It was, however, by no means a united and homogeneous bureaucracy. Over the course of its turbulent history, its various branches produced divergent and often contradictory practices and ideologies.

An early moment of transition in NAD thinking was brought about by the advent of settler rule and the turn to territorial segregation in the 1920s. At this time, the NAD began to move towards a more or less overt policy of relying on chiefs and headmen to perform administrative tasks. The move marked a significant shift from earlier visions of a 'detribalised', proletarian African population in which 'tribal communalism' was to be replaced with the 'individualism of the European'.[3] 'Tribal leadership' was codified for the first time in the 1927 Native Affairs Act. The Act was intended to reinforce mechanisms for maintaining control and extracting tax and labour, duties for which increasingly desk-bound Native Commissioners (NCs) no longer had time.[4] Judicial powers were gradually devolved to chiefs, steps which were justified in terms of 'restoring' traditional powers, but which were also a direct response to growing expressions of African opposition.[5]

NCs increasingly (though unevenly) used traditionalist arguments both to explain and to discredit opposition. African grievances needed to be cast as illegitimate. Thus NCs routinely detailed strings of complaints – discriminatory pricing, forced labour, taxation, evictions – and in the same breath described political opposition as the work of malcontents and agitators, almost invariably of urban origin, who only managed to gain support because of the disruption of 'tribal life'. They increasingly sought a remedy for opposition in what was assumed, often mistakenly, to be the conservative influence of chiefs and headmen. Though not elaborated in the scientific terms it would later assume, this marked the beginning of an ideological discourse around chieftaincy which would provide an important basis for NCs' claims to expertise within the NAD, and to legitimacy with regard to Africans.

Parallel to these developments, the NAD turned its energies to intervention in the ways in which Africans lived and farmed within the reserves, a focus again shaped by territorial segregation. 'Technical development', as it was later called, encompassed four main policies: demonstration, centralisation, destocking and, finally, the Native Land Husbandry Act (NLHA).[6] Each of these policies was justified in very different ways, but all were premised on a set of beliefs which assumed the superiority of western culture and science.

E. D. Alvord, an American missionary and agriculturalist, played the central role in initiating interventions in African agriculture. In 1926, he was appointed as the first 'Agriculturalist for the Instruction of Natives' in the NAD. He initiated the demonstration policy at a time of official concern over African production methods, in part a result of the cost of famine relief in the early 1920s. But Alvord's own motives were diverse. He saw agricultural change as part of a 'wider civilizing package' incorporating Christian conversion. Demonstration would, he argued, win 'converts' to 'modern scientific agriculture', defined at this point as intensive permanent cultivation, crop rotation and the use of fertiliser and improved seeds.[7]

Centralisation was a more ambitious policy requiring the reorganisation of settlements into linear villages along watersheds such that they would separate grazing areas from fields. Like demonstration, centralisation was part of a civilising agenda propounded by Alvord the missionary as much as Alvord the agriculturalist. It also had an administrative appeal: some NCs saw it as an ideal way of bringing dispersed populations under their control.[8] Under pressure from white farmers, the purpose (if not the form) of centralisation was recast in the context of the 1930s' Depression: it was no longer described as a measure designed to promote African 'development', but rather as a conservation measure and means of 'squeezing' more Africans into the reserves.[9]

Alvord's methods were superseded in the 1940s by a far more coercive and interventionist ethic, promoted by a rapidly expanding corps of technical officials within the NAD, and justified by a growing and region-wide alarm over conservation.[10] In the early 1940s, the passage of the Natural Resources Act, the establishment of the Natural Resources Board (NRB), and the Report of the Native Production and Trade (or Godlonton) Commission, laid the foundation for a new regime. The NRB focused its attention on 'overstocking' and, in 1943, compulsory destocking regulations were passed. More widely, in an effort to halt what was described as the 'wholesale rapid destruction [of the reserves] due to bad tillage and overstocking', a bevy of soil conservation, agricultural, community and livestock demonstrators, along with pasture officers, land development officers and lands inspectors, roamed the reserves armed with a growing array of coercive powers.[11] As one NC put it, these officials acted as 'virtual policemen', bent on prosecuting 'agricultural crimes'.[12] African farming methods were, in effect, criminalised.[13]

The Godlonton Commission set forth even more ambitious plans for intervention. It called for:

compulsory planned production whereby a statutory body should be empowered to direct what crops, acreage and areas should be planted and what livestock should be kept, to enforce good husbandry conditions and to control the distribution and marketing of consequent products.[14]

Michael Drinkwater has aptly described the tenor of this shift:

> The secretary of the Godlonton Commission was Arthur Pendered. If Alvord can reasonably be cast as the architect and guide of the policies initiated in the first half of the technical development period, Pendered was the key figure in the second half. Alvord was an ex-missionary, Pendered a technocrat, and this contrast is symbolic of the shift between the pre-Second World War and post-war colonial times. Alvord was an active agricultural evangelist in the rural areas; Pendered remained a remote manipulator.[15]

It was in this context that the NLHA was conceived.

The Native Land Husbandry Act

In 1951, the NLHA was promulgated amidst an economic boom and overwhelming confidence in the prospects for state-planned modernisation. The Act marked the height of Rhodesian technical development and would significantly transform African – as well as white – politics. Montague Yudelman described the Act as 'one of the most far-reaching land reform measures in Africa';[16] African nationalists called it their 'best recruiter';[17] without a doubt, it provoked 'the most violent outbreaks of rural opposition to colonial rule since the First Chimurenga of 1896/97'.[18]

Changes in the Rhodesian economy after World War II created the context for the rise of technical development and the dominance of technical officials within Native Affairs. Growth in secondary industry and a boom in tobacco production created severe labour shortages in the formal sector of the Rhodesian economy.[19] At the same time, massive evictions from 'European' land consequent on white immigration increased pressures within the African reserves, and heightened the already widespread concern over conservation. For many NAD officials, these developments and concerns seemed to confirm beyond doubt both the possibility and the necessity of a radical restructuring of African participation in the urban and rural economies.

The NLHA was a product of this context, and marked the realisation of many of the recommendations of the Godlonton Commission of 1944. Its passage was preceded by a major revamping of the bureaucracies concerned with administering and developing the reserves. The so-called specialist branches – Education, Labour, Agriculture, Accountancy, Native Engineering, Native Marketing – were, with District Administration, cast into a single superstructure, the Division of Native Affairs. Arthur Pendered, previously secretary of the Godlonton Commission, secured the key post of Under Secretary of the Division from which 'he acted as a spokesman for the "progressive" technical branches rather than for the "conservative" administrative branch'.[20] The technical branches rapidly expanded and, as Drinkwater writes, 'the belief in the superiority of formal scientific knowledge was institutionalized in dogmatic form'.[21]

In this context, ideas about 'traditional rule' and 'tribal structure' elaborated by the administrative cadres took a back seat. A bold new scientific ethos now held sway. As one of the most sensitive contemporary commentators on this period, J. F. Holleman, writes, technical officials,

> had gathered hard facts, were building up a vast organization of specially trained people, and were able to present concrete plans... Against this elan the administrators could do little better than advise caution, and speak vaguely about the dangers of disturbing the social system and of the desirability of consulting chiefs and headmen. At no stage during the crucial period of planning did they produce a comprehensive statement of what this social system actually involved.

NCs were no longer the 'principal custodians of tribal life' – they were now the ill-prepared 'captains of its social and economic transition into the modern world'.[22] As the Chief Native Commissioner (CNC) commented in 1951, 'There is no stopping these agriculturalists and economists. We want economic improvement in the Reserves, and these chaps seem to have all the answers.'[23]

The Act in fact placed administrators in a difficult position. Technical officials were little concerned with the practice of administration or the practicalities of enforcement. The Godlonton Commission had merely called for the appointment of younger chiefs and their education in 'progressive' attitudes and scientific agriculture such that, 'better discipline and greater sense of responsibility will develop, and agriculture, animal husbandry and conservation of natural resources will... be greatly benefited'.[24] Increases in chiefs' powers as implementors of a policy that would, in the end, undermine their position, and a belated effort to give African Councils a new lease on life, was all that was forthcoming.

Of what did the NLHA consist? To a large extent, it simply extended and enforced earlier policies with more punitive powers: the precepts of centralisation still shaped settlement patterns and the division of land between individually farmed arable blocks and communal grazing areas; destocking and grazing schemes remained the principal means of stock management; physical conservation works and the methods of agricultural intensification developed by Alvord continued to dominate interventions into arable production. However, the NLHA was revolutionary in other respects: unlike previous policies, it was not intended to 'squeeze' more Africans into the reserves but to put a stop to labour migration between the reserves and urban areas entirely, and to bring a halt to further settlement in the reserves by issuing saleable land and stock rights to a permanently limited number of African farmers. That is, it was intended to transform African participation in the economy, making full-time farmers or workers of all, and it was intended to alter radically African land tenure from 'communal' to individual, at least with regard to arable land.

In keeping with the development orthodoxy of the period, officials' justifications for the NLHA rested heavily on the argument that it would increase economic efficiency. Arthur Pendered wrote,

> The time has now come when all indigenous natives can no longer continue to maintain a dual existence as part-time employment in the European areas and part-time farming in the Native Reserves for, apart from its impossibility, it does not conduce to efficiency in either area, nor can the economy of the colony afford to offer satisfactory conditions in both areas for the dual mode of life.[25]

Those excluded from the reserves would provide a stable workforce for industry and white agriculture. In the reserves, limiting the number of farmers and giving them secure tenure was intended to promote investment and allow the enforcement of conservation measures and recommended agricultural methods. Technical officials expected a direct and 'almost immediate beneficial result' in productivity and conservation to follow from the NLHA's (in theory) secure tenure.[26]

Administrators were themselves captivated by this logic, their long-standing defence of the merits of custom and tradition notwithstanding. As the CNC bluntly put it:

> The communal system and capitalism are incompatible and cannot flourish within the same economy. The communal system of land tenure hangs as a dead weight upon the Native and frustrates all his endeavours to attain Western standards ... It retards his transition from a subsistence to a money economy, and is out of keeping with our declared purpose in Southern Rhodesia, which is to develop the resources of the country to the utmost.[27]

The idea of making a 'final allocation of land' in the reserves such that, 'the Native will either become a peasant farmer only' or 'an industrialised worker with his tentacles pulled out of the soil' also appealed to many administrators for the reason that it promised to cap the unceasing demands for land from Africans faced with eviction, a source of both great annoyance and paternal concern to NCs.[28]

The NLHA thus directly repudiated 'customary' and 'communal' rights to land in favour of individual right holders and 'secular state power' – that is, the government officials who monitored land use and land transfers.[29] Yudelman argues that, 'It was the intent of the act that the intricate network of social and tribal customs regarding land use and land transfer would give way to the marketplace.'[30] But this commitment to the 'marketplace' was tempered by a host of restrictions: only those who farmed and owned stock at the time of the Act's implementation were eligible for rights; land rights could not be used as collateral against a loan; the size of arable allocations and number of stock rights was limited in line with technical estimates of carrying capacities, while accumulation was limited to three times the 'standard holding' designated for each reserve. Conservationist concerns were used to justify an array of punitive measures to enforce 'good husbandry' and provide labour for conservation works.[31]

The heavy state regulation of the reserve 'marketplace' and agricultural production under the Act reflected officials' belief in the need for close state control over the process through which Africans' 'backward and inefficient' farming methods were to be modernised. This was 'innovatory paternalism': if officials were willing to concede that Africans could be modernised, they were certainly not willing to let them do it themselves.[32] What was needed was the collection and analysis of data by a huge corps of experts on an immense scale and at huge cost: fifteen million pounds were spent between 1950 and 1958, a large part of which was drawn from levies on marketed African produce.[33] Employing recent advances in aerial photography, an initial mapping and land classification process was to be undertaken in each reserve, and the need for water sources assessed.[34] Land Development Officers with African staff were then to take a census of farmers, arable land and stock. Detailed land classification and conservation plans were to be drawn up by specialists and an Assessment Committee was then charged with recommending standard field sizes and stocking rates. Destocking would follow, and farming and grazing rights would be registered. The registered holder was then legally responsible for the construction of conservation works within one year. After all this had taken place, a final plan for land use was made, agricultural extension resumed, and a second round of aerial photography undertaken.

These costly and elaborate procedures left chiefs, headmen and administrators with much reduced control and influence. The administrative cadres did not, however, seek to do more than tinker at the margins: in 1951, the year the NLHA was promulgated, they undertook a project of 'rationalising' chieftaincy, reducing the numbers of chiefs and increasing their subsidies.[35] A Provincial Assembly for chiefs was also created. As the NLHA was implemented, administrators did not question its validity but they did begin to ask for further measures to bolster the position of chiefs, arguing that the 'onerous and unpleasant' duties which chiefs and headmen had to carry out under the NLHA meant their loyalty was severely strained.[36] As nationalist activity burgeoned in response to the Act, they regarded the position of chiefs with increasing concern. As the Minister of Native Affairs wrote in 1957:

> It is well known that many leading speakers of the Southern Rhodesian African National Congress are against such things as the Land Husbandry Act, and are sowing seeds of discord amongst the Africans in the Reserves. From the meetings already held, it would appear that the exploitation of imaginary grievances, e.g. destocking and the allocation of land, is the method employed to win support ... We must obtain and maintain, through the Chiefs, administrative assistance and political stability.[37]

The Minister argued that, 'If the Government fails to range the chiefs on its side then either the chiefs will surrender to the demogogues [sic], or the people, denied a strong lead by the chiefs, will be at the mercy of every soap box orator.'[38] As in previous periods chiefs were cast as 'a steadying conservative influence among their people'.

It was partly in response to this growing strain that a shift within the NAD began to take place. Initially, as suggested above, technical officials exercised unprecedented control. With the help of an alarmist report on the state of conservation by the Natural Resources Board in 1954, they vastly increased the speed of the Land Husbandry Act's implementation.[39] But only a few years later, as growth in the national economy slowed and nationalist mobilisation boomed, the confidence and dominance of the technical branches waned. Technical officials themselves began to argue that the NLHA's assumptions and methods of implementation were flawed.

The NLHA's flaws have been discussed at length, in both academic and official circles.[40] Here I touch on only a few of the problems relating to productivity and land shortage. The 1955 NLHA Five Year Plan had projected large increases in productivity: five years after implementation, crop output was expected to increase by 50 per cent; cattle output was expected to increase by the same amount in eight years.[41] These goals were not realised for a number of reasons. Agriculturalists had projected increased yields on the basis of the yields allegedly achieved by the less than 1 per cent of farmers who were demonstration plot holders.[42] There were, in addition, serious constraints on establishing the 'economic unit' on which the yields were to be realised as there was nowhere near enough land in the reserves for all those eligible to have a full 'economic holding'. The Quinton Committee noted in 1960 that there was land available for only 235,000 of the eligible 346,000 farmers.[43] Officials at first sought to accommodate more farmers by reducing the size of the 'economic holding', undermining its already questionable validity.[44] They then sought to reduce the numbers of claimants to rights by speeding the pace of registration, thus arbitrarily excluding a good many farmers.[45] But even the accidental and deliberate exclusion of vast numbers of farmers could not solve the problem of land shortage. And the fact of their exclusion greatly undermined, rather than increased, security of tenure – one of the key goals of the Act.[46]

These were not the only problems: there were also severe 'imbalances' between arable and stock holdings. Stock was required to maintain fertility in arable land under the agriculturalists' model, but often farmers had too few or no stock, and this in areas that faced severe destocking in line with pasture officers' assessments of carrying capacity.[47] Serious difficulties also emerged in the process of planning due to speed and the poor training of field staff. Agriculturalists complained that the Technical Block responsible for training simply could not cope: the result would inevitably be 'chaos'.[48] In fact, all agricultural staff were under extreme pressure. Virtually all staff had been diverted from their regular duties to NLHA implementation. The Director of Native Agriculture noted in 1958 that extension work had ground to a halt, that 'Every member of every Provincial Technical Block has virtually been turned into a Land Use Planning Officer', that Soil Conservation Officers were under 'considerable strain'.[49] In practice, shortcuts and abrogations of technical prescriptions were the norm in the process of implementation, and Africans took full advantage of these weaknesses to evade prescriptive demands.[50]

If all this were not enough, it became clear in the second half of the 1950s that the industrial and white farming economy simply could not absorb those denied land, thereby creating a massive category of 'landless unemployed'.[51] It was also clear that wages, housing and welfare provision in the urban areas were far from adequate for the envisioned working class, and that the extent to which agricultural production relied on migrant labour remittances had been sorely underestimated.[52]

These pressures and problems were heatedly debated in the national forum of the NLHA Standing Committee, and often led to clashes between administrators and technical officials. Destocking was perhaps most controversial of all. NCs argued that it was causing tremendous administrative difficulties, 'political restlessness', and undermining their relationship with chiefs and headmen. They therefore asked for destocking to be carried out over longer periods. The Director of Native Agriculture responded by blaming NCs for relaxing controls on cattle in advance of the NLHA's implementation, thus creating the very problem that they now faced. NCs were able to extract only a minimal compromise. Small changes could be made in the timing and method of implementation, but only, as the Director of Native Agriculture put it, 'in deference' to the Act, that is, in order to ease its continued implementation. In practice, NCs routinely abrogated technical prescriptions or found them impossible to enforce.[53]

Perhaps the most devastating critiques of the NLHA came from economists themselves. Pending, or as a result of, economic evaluations, the Act was suspended on irrigation schemes in 1956 and in high rainfall areas in 1957. In 1959, Native Affairs Economist Arthur Hunt carried out an evaluation of low rainfall areas. He questioned the assumption that these areas were primarily cattle-producing regions, as well as both the efficiency and the viability of production on the NLHA model.[54] In an economic report which concerned the country as a whole, Dr S. M. Makings, of the Department of Economics and Marketing, called for 'close scrutiny' of the NLHA and summed up the increasingly pervasive doubts about the Act's most basic assumptions:

> The Act is based on the major principle of individual incentive and responsibility as the outcome of individual ownership and the latter is given status as perhaps the chief factor in forward development. But if in fact there is little prospect for individual economic holdings except in limited areas it is very important that forward planning should take full cognisance of this position. It would be better to found thriving communal systems in which there were necessary restraints in the common welfare rather than a patchwork of individual holdings forever battling with abject poverty as the price of independence.[55]

These profound doubts were underlined by the findings of the Central Statistical Office's *Sample Survey of African Agriculture.* Undertaken in 1959–60, the survey showed that the NLHA had had a negative effect on yields and crop diversity in most regions. Though the results were withheld from publication until 1962, they were circulated among officials.[56]

There were thus, from 1957, cutting critiques of the NLHA's most basic assumptions from within the technical branches which had so recently championed its promulgation. However, the final blow to the Act came less from technical doubts than from the political crisis which the Act engendered. This was important in reprieving the technical sciences from condemnation, as well as in allowing the administrative cadres to formulate their own, scientifically justified, alternative. The dramatic shift in the sciences of development was in the end less a product of scientific debate itself than of political circumstance and bureaucratic rivalry.

Political crisis and the human factor

From 1959, the Rhodesian government increasingly turned to repressive measures in response to growing nationalist activism. A State of Emergency was declared between February and April 1959. Immediately thereafter, parliament passed the Unlawful Organizations Act and the Preventive Detention Act. The ANC was banned under the former while the latter was used to detain over 300 nationalist leaders, mainly in rural areas. The 1959 Native Affairs Amendment Act made it a crime for any African to say or do anything 'likely to undermine the authority' of government officials, chiefs or headmen and prohibited meetings of twelve or more in the reserves save with the NC's approval. In 1960, the Law and Order (Maintenance) Act and the Emergency Powers Act were passed, both of which greatly increased the powers of the executive and security branches of government. Between 1958 and 1962, the personnel of the British South Africa Police and Ministries of Native Affairs and Justice nearly doubled while the number of police stations rose from 102 to 134.[57]

In the rural areas, the degree to which civil servants, notably administrators, felt directly threatened by opposition to the NLHA was clearly expressed in March 1961 by the Native Affairs Advisory Board (NAAB).[58] Members saw the last two years of resistance as 'nothing less than a challenge to the right of the Government to exercise any control'.[59] CNC S. E. Morris warned, 'there will be bloodshed' if destocking continued.[60] The Board cited growing instances of resistance to the Act:

These have taken various forms such as opposition to veterinary measures, creating disturbances at cattle sales, breaking up Land Development Officers' extension meetings, assaults and threats of assaults to persons in authority, down to direct opposition to the application of the Native Land Husbandry Act and the deliberate flouting of its provisions after application.

The Board concluded on an apocalyptic note:

> *The Important point* in all this is the extent to which this criticism and opposition to what is Government policy is undermining and obstructing the administration of this policy and the whole maintenance of law and order. This can and will, unless it is strongly combatted and counteracted, bring about the breakdown of administration and maintenance of law and order. This could all happen in a relatively short time. *This all important fact must be brought to the attention of the Government in the plainest terms.*
>
> The most urgent and dangerous aspect of this overall position is the determined and by no means unsuccessful campaign to penetrate the rural areas with planned political agitation in which every possible grievance, particularly in relation to land and stock, are [sic] exploited to the full, emotions whipped to the point of violence and the fullest use made of intimidatory practices.[61]

In 1960 and 1961, strikes, demonstrations and riots in the urban areas mirrored the loss of control in the rural areas.[62]

One of the Board's principal concerns was over the position in which NCs had been placed by the NLHA, a position which members considered manifestly unfair because, 'it is Government and Government policy that is being attacked, the Native Department is simply the most convenient and obvious target'.[63] NCs were subject to 'insults, indignities and even the possibility of physical assault'.[64] 'The most significant aspect of this agitation,' one contributor wrote, was that it engendered

> strong feelings of hostility towards the native commissioner, who is looked upon by the people as the person who decides whether or not a person may have land or stock rights. The most pressing and immediate problem is to direct these feelings of hostility away from the native commissioner.[65]

Just where such feelings should be directed went unspecified.

The attacks on NCs brought to a head the strains between the technical and administrative branches of Native Affairs. Arthur Pendered summarised the prevailing views in 1961.[66] Administrators felt that:

The professional and technical staff appear to want to proceed from an advisory to a directive capacity, but, on the other hand, do not want to accept responsibility for distasteful tasks, e.g. destocking, Native Land Husbandry Control, etc.

Some NCs felt, 'they are being overwhelmed by technical experts who have virtually taken over district planning and development policy and that there is a danger of district administrative control passing out of their hands'. Technical officials, on the other hand, held that, 'It is wrong that a technical service should be directly controlled by an administrative head.' They wanted 'real responsibility' and independence. Others once again questioned the NLHA itself, drawing attention to its technical flaws, hurried implementation and the complete neglect of extension work. Technical officials also noted the difficult position of young and inexperienced Land Development Officers, and argued that 'inspection, control and so-called quasi-police work in connection with [NLHA] control, including destocking and veterinary control is incompatible with gaining and retaining the goodwill and cooperation of the African farmer'. Neither group wanted to shoulder the increasingly dangerous burden of enforcing the Act.

Thus, in early 1961, senior officials of the Ministry of Native Affairs widely perceived the NLHA as a threat to their officers, as well as the government's very ability to rule. The long-standing tension between technical and administrative officials was pushed to the fore.

In the context of this panic over African opposition, the negotiation of a new constitutional relationship with Britain, and moves to majority rule in the Central African Federation, the Rhodesian government established a series of official commissions to assess agricultural and administrative policy. The commissions were strongly influenced by imported 'experts' and assumed that Rhodesia was moving towards, first, a gradual desegregation of land and, second, the adoption of a multi-racial policy of community development – a step first mooted by CNC S. E. Morris in 1959 and formally adopted by Edgar Whitehead's government in June 1962.[67] Thus the Mangwende, Robinson, Paterson and Phillips Commissions sat and reported on all aspects of administrative, judicial, agricultural and economic policy in the early 1960s. In addition, a community development adviser, Dr James Green, was engaged under the auspices of the United States Agency for International Development (USAID). Green was 'a consultant sociologist with international experience in community development'. He produced his own reports as well as significantly influencing the commissions.[68]

In brief, the commissions' recommendations resulted in the creation of a new Ministry of Internal Affairs (MIA) to replace the Native Affairs Division. Technical and judicial services were removed from the direct control of Native Affairs while the new MIA was left in the key role of coordinating services and running local government under the rubric of community development. The reorganisations were implemented and modified over the 1960s with varying effects but with an overriding tendency towards returning the policy initiative with regard to development in the reserves to the administrative branch.[69]

The return to prominence of administrative concerns was instrumental in redefining the causes of African opposition to the NLHA in terms of what were called 'human factors' – rather than the Act's by then widely noted scientific flaws. The 'human factor' gained currency in the official discourse of both the technical and administrative branches in the early 1960s, partly drawing on James Green's use of the term. Thus implementation of the NLHA had to be slowed down so as to 'deal with human factors'; resistance resulted from the neglect of 'human material'; technical development had produced 'a kind of human sand held together with little or nothing'.[70] In his important review of technical development policies in 1961, CNC Morris wrote:

> What is the common factor that has produced results so different from the confident hopes of those who produced these schemes and threw finances, personnel and technical specialists into the battle to save the soil and increase production? As we now see it, the simple explanation of a vastly complex problem is that a barrier of human beings lay between technical knowledge and the soil. And that barrier comprises not a multitude of individuals waiting to welcome benefits, but a cultural organisation with its own structure, organisation, patterns of thinking, feeling and acting ... The problem is not a technical one, nor is the crux of it the extent to which capital, management and labour are there to develop the land, as the economist would have it. It seems clear that the more a technical approach is adopted ... the greater becomes the cultural resistance. We must see it as a human problem.[71]

This soul-searching report represented a direct attack on technical dominance within Native Affairs, and laid the groundwork for the application of a new discourse of human science to the long-standing practices of administrative traditionalism. If resistance was 'cultural' and the problem 'a barrier of human beings', then the answer lay not with technical calculation but in administrative solutions provided by community development and backed by a different band of experts – sociologists and NCs. The community development policy, as formulated by Green, fit well with existing conceptions regarding African society within the administrative cadres. Green found that the best basis on which to 'delineate' communities to which responsibility for services could be transferred was the 'judicial function', a function 'invariably' exercised by chiefs and headmen. According to the community development philosophy, technical officers' attempts to create 'economic units, based on a balance of arable and grazing land' were 'sociologically unwise' and hence 'doomed to failure'.[72]

Green's findings and the beliefs of the administrative branch were further reinforced by the Whitehead government's need for an alternative voice to that of the nationalists. To this end, the government set about transforming the chiefs' official role by recruiting them into national politics. The chiefs' first act was to deliver a badly needed measure of African approval for the 1961 constitution at a government 'indaba'. In return, parliament passed the Council of Chiefs and Provincial Assemblies Act, establishing a prestigious forum for consultation with selected chiefs; chiefs' salaries were raised, and further concessions were promised. At the same time, increasingly repressive measures were enacted against nationalist activists, thus effectively bringing the 'open' nationalist period to a close.

At all levels of government there was now a strong disposition to turn to chiefs – rather than a more democratic system potentially dominated by nationalists and lacking the 'natural' cohesion Green sought for community development. Communal not individual rights in both economic and political spheres were emphasised. The debate within the technical branches over the validity of their recommendations was displaced in this process, a development which suited many of the technical officers themselves, especially those who had expressed doubts over their recommendations, suffered physical attacks, or desired a return to the less fraught business of agricultural extension.

Acceptance of an administrative solution to resistance nevertheless had to be given scientific validity. There thus followed a huge programme of collecting information about 'traditional structures': once again, the exercise and legitimation of power depended on the collection and systematic ordering of knowledge. This transition was not to be a smooth one: technical and community development policies coexisted during an awkward interim period. The Robinson Commission's Working Party D is illustrative in this regard.[73] Dominated by administrative officers and chaired by the chief architect of the NLHA, Arthur Pendered, it combined, in a highly unsuccessful way, the concerns of both community and technical development.[74] The Commission's brief was straightforward: 'to propose measures designed to fill a vacuum in the reserves and to retain the loyalty of the chiefs'.[75]

TECHNICAL DEVELOPMENT AND THE HUMAN FACTOR

In making its recommendations, the Working Party was strongly influenced by its consultations with chiefs and administrators and by an overriding sense of the urgency with which the 'landless problem' had to be addressed. At the same time, it continued to draw on technical advice and methods.

Working Party D found that chiefs' loyalty was 'severely strained' by 'pressure being brought upon them by their followers as the result of political agitation or by what they consider to be legitimate grievances'. A bargain urgently needed to be struck: some concessions on land were deemed necessary and the status of chiefs needed to be bolstered with powers over land and in courts. A failure to undertake these measures would cause 'a breakdown of good Government, administration and extension'. In the short term, the 'political menace' and 'major obstacle to efficient administration' posed by the 'landless' needed to be met by allowing chiefs to allocate land immediately. Hastily demarcated areas in grazing land were 'placed at the disposal of the tribal authorities' largely on the basis of multiplying the 'known number of landless' by the 'subsistence acreage per family for that region'.[76]

Under this thin veneer of technical calculation, the Working Party gave free rein to chiefs: they were to act on the basis of 'tribal law and custom'; and they were found to have 'a very lively and detailed recollection of the tribal authorities' powers and responsibilities in regard to land allocation'. The Working Party thus sought the 'resuscitation' of 'Tribal Land Authorities', and adopted an extraordinarily laissez-faire attitude: 'those who opt for the traditional tribal system in land allocation must be assumed to know what they are doing and to be prepared to accept the consequences'.[77] Chiefs were, however, often all too aware of the consequences, and many refused to cooperate in the absence of land redistribution.[78] The Working Party went on explicitly to reject individual tenure in the reserves, thus scuppering the basis on which the Quinton Report had envisaged the gradual desegregation of land, through the conversion of all categories to freehold status.[79] The reserves were to remain a traditional bastion, bound by custom. The Working Party argued,

> We may even have to return for the time being to the old concept of the tribal lands as the 'security pool' of the tribe and 'reservoir of labour' until we can fill in some of these gaps (e.g. social security, adequate industrialisation, administrative-development approach, etc.), which exist in our society and economy as compared to that of a modern state.

But if Rhodesia was clearly not ready for modernisation, this did not explain the reasons for the failure of the NLHA: the Working Party interpreted resistance to the Act, and justified the 'return' to 'tribal tenure', in terms of Africans' 'spiritual' ties to land and cattle, arguing that these attachments were 'quite impervious to logical or other argument'.[80] This view was cited repeatedly by the commissions of the early 1960s: though the NLHA's wide-ranging economic and technical flaws were repeatedly identified, African objections were not in any way related to these deficiencies.[81] There was a 'much deeper reason', that is, 'the clash between the objectives of the Act and the traditional beliefs that the *Chiefs spiritually are the land*, that the *individual cannot have rights to ownership of the land*, and that *cattle, possessing a spiritual significance far transcending their intrinsic value, cannot be treated merely as animals providing services or for disposal for sale*'.[82] This view, though not new, was given a powerful new validity, sanctioned by the experts who sat on the commissions and working parties, and bolstered with reference to anthropological insight.

Acting on these assumptions, the Working Party set out:

(1) to determine what was the traditional Native Authority in respect of the land according to tribal law and custom, and what were its methods and powers; (2) to list the ways in which Government policy and legal enactment has altered or eliminated these powers; (3) to consider whether and in what form this authority should be restored and what powers it should be given to make it effective in the changed circumstances of today.[83]

In its deliberations, administrators played an important role. The 'overwhelming majority' were found to believe that 'traditional tribal authority is by no means dead, that its force and continuing viability reasserts itself strongly at every turn and that certainly for some time to come it has a very important part to play'. With no sense of irony, they stressed that chiefs had been a great help in implementing the NLHA![84]

Working Party D also argued that the 'tribal authority' would provide 'cohesion' in line with the community development goal to 'build upon and to develop from existing institutions'.[85] The language of Green's community development policy was used to cast the turn away from technical development in the positive light of a new, more enlightened administrative philosophy, sanctioned by sociological and anthropological insight, and given an earthy credibility by administrators' testimony to their hands-on experience with 'tribal custom'.

By the end of 1961, many of the policy decisions crucial to undermining state-directed agrarian change had been taken. The final blow came in 1962 when Working Party D commissioned a countrywide survey in a last and ultimately doomed attempt to dress up chiefs' land allocations in the trappings of technical development. Very much in the style of the NLHA's implementation, Pendered stated that the survey 'obviously calls for the coordination of a vast amount of data and a series of recommendations in respect of each district'.[86] Technical officials hurriedly toured the country, assessing 'ecological risk' on the basis of the 'Annual Cropping Acreage' and the 'Recommended Livestock Equivalent' for each of 210 areas.[87] The former was calculated by reducing the 'Potential Arable' land (itself determined by excluding all land with over a 12 per cent slope or with less than ten inches of top soil) by factors which determined the amount of land on which productivity could be sustained under a given – but unclearly specified – farming system.[88] In order to create what it called 'room for manoeuvre', the Working Party, in negotiation with technical officers, agreed to allow up to 30 per cent overstocking. The Working Party assumed that a judicious measure of overstocking, and the use of what were felt to be more sophisticated technical methods, would allow increases in carrying capacities and arable land allocation. It was to be sorely disappointed. The findings and the fate of the technical surveys illustrated both the lack of consensus among technical officials and the pride of place which political considerations and the 'human factor' had come to assume.

Makoni District offers an extreme but not unrepresentative case of the problems which the survey findings encountered in higher rainfall regions. The new stock-carrying capacity assessments were drastically lower than previous assessments, leaving 'no room whatsoever to manoeuvre'. Though 'technically irrefutable', according to the Working Party, the assessments were strongly challenged by the NC and left the Provincial Agriculturalist, himself a pasture expert, 'somewhat taken aback'. The NC stated his case vehemently, and very much in keeping with the spirit of the day:

> As I have so often pointed out, the erosion of the soil is one thing, but with certain measures it can be healed, whereas the erosion of the goodwill of the Africans, which is likely to be caused by lack of arable land is an entirely different matter, and, so far as I know, once lost there is no remedy available to regain it. It would appear to me, therefore, that whatever the technical considerations are, the human ones completely outweigh them.[89]

In lower rainfall areas, where the stress lay on pastoral production, there were if anything more severe obstacles. No agreement could be reached between agricultural and administrative officers. The Working

Party thus established an arbitrary measure of acceptable usage of arable land which it freely admitted did not have 'any technical merits'. The same problem arose with regard to stock: new technical assessments often lowered existing assessments of carrying capacity; administrators and technical officials proved unable to reach agreement. Thus the Working Party decided to disregard all previous stocking policy, resolving instead that, 'Recommended numbers must be regarded purely as a guide line because (a) the number of stock is constantly changing, (b) the carrying capacity differs from one year to another.'[90] In practice, throughout the country, technical officers' often hugely contradictory recommendations were rejected at the behest of NCs and in the name of 'practical politics' and 'human factors' – but not because they were deemed invalid.

In its efforts to rebuild rural administration, Working Party D set in motion processes which would undermine many of the long-sacrosanct tenets of Rhodesian technical planning. The commitment to mechanical conservation was, however, rigorously defended as a 'prime duty of the Government'. The community development policy explicitly designated conservation as a 'national issue' which should not fall under local control.[91] Conservation measures, notably the hated contour ridges, were enforced well into the period of the guerrilla war.

In sum, within the space of a few turbulent years, a new orthodoxy had been entrenched and given scientific validity: it rejected the possibility of 'modernising' the reserves on the grounds that Africans were too steeped in tradition to accept the rationality of western science, individual tenure, private property and modern markets. The turn to chiefs brought together the traditionalist beliefs of administrators and their desire to maintain 'order', the 'expert' advice of community development advisers seeking 'natural' communities, and white politicians' need for allies other than the nationalist leadership. It entrenched territorial segregation in the process, by underlining the essential difference between African and European, and confirming the essential superiority of European over African ways.

The full implementation of this new orthodoxy took some years. The focus on community development, natural communities and tribal authorities, needed to be validated by a massive project of information collection. Thus, between 1963 and 1967, all chieftaincies within Rhodesia were 'delineated'. Internal Affairs' researchers travelled the length and breadth of Rhodesia, collecting chiefly histories, establishing the boundaries of 'traditional' communities, and noting down 'felt needs'. Administrators went to often absurd lengths to identify 'legitimate' chiefs on the basis of traditional succession procedures. They became victims to the 'hypnotic attraction of "inherent" customary authority', to the 'cult of tribal custom'.[92] When this phase of community development failed to deliver either conservation (which remained the great obsession of the technical branches) or nationalist quiescence, Internal Affairs argued that the problem came not from 'tribal structure or organization as such, but from using incorrect levels of organization'.[93] Smaller social units needed to be identified. Thus, in 1973, another great delineation exercise was launched, this time in search of village-level communities, an endeavour which eventually succumbed to the advancing guerrilla war.[94]

Conclusion

This period in Native Affairs' thinking about the development of the reserves can be seen as a key moment in the displacement of one scientific discourse and practice with another. The most important impetus behind such change was the political crisis provoked by the NLHA's implementation – not the scientific critique mounted from within the technical branches themselves. In the search for a solution, vague views about tradition and tribal structure held within the administrative branch were accorded a new validity by the work of commissions and international experts. Sociology and the 'human factor' displaced economics and agricultural science. The way in which this transition occurred had important consequences. Technical development was not itself blamed for the demise of the NLHA – rather, the newly scientific analysis of African society provided a neatly self-contained explanation for both resistance and the NLHA's failure, as well as a basis for the resolution of the quandary. Modernisation on the NLHA's model was valid, and was never explicitly renounced: Africans, and African society, simply were not ready for it. In the heyday of Rhodesian Front racism which followed this moment, the assumptions about the cultural 'otherness' of Africans established by community development advisers and expert commissions alike would be used to new effect. In the post-independence period, the precepts of technical development would be revived once again.

Notes

1 I owe many thanks to a host of people and institutions who helped during the research for and the writing of this chapter, but shall limit myself to thanking Terence Ranger who supervised the thesis on which this piece draws, the staff of the National Archives of Zimbabwe who generously allowed me access to as yet uncatalogued material, and to Saul Dubow for his insightful comments.
2 D. A. Low and J. Lonsdale, 'Introduction: Towards the new order 1945–1963', in D. A. Low and A. Smith (eds), *History of East Africa*, vol. III (Oxford, 1976), pp. 12–16.

3 M. C. Steele, 'The foundations of a "Native" policy in Southern Rhodesia, 1923-33', Ph.D. dissertation, Simon Fraser University, 1972, pp. 80-2.
4 J. F. Holleman, *Chief, Council and Commissioner* (London, 1969), pp. 35-43; Steele, 'The foundations of a "Native" policy', pp. 80-6.
5 R. Smith, '"There and back again": The integration of unofficial courts in Southern Rhodesia', SOAS African History Seminar, London, MS, 1992, p. 5.
6 See M. Drinkwater, 'The state and agrarian change in Zimbabwe's communal areas: An application of critical theory', Ph.D. dissertation, University of East Anglia, 1988, p. 287. On debates over the establishment of departments concerned with 'development', see Steele, 'The foundations of a "Native" policy', pp. 306-51; D. J. Murray, *The Governmental System in Southern Rhodesia* (Oxford, 1970), pp. 238-9.
7 J. McGregor, 'Woodland resources: Ecology, policy and ideology: An historical case study of woodland use in Shurugwi communal area, Zimbabwe', Ph.D. dissertation, Loughborough University, 1991, p. 84. On the development of Alvord's ideas, see also Steele, 'The foundations of a "Native" policy', pp. 316-22; M. Drinkwater, 'Technical development and peasant impoverishment: land use policy in Zimbabwe's Midlands province', *Journal of Southern African Studies*, 15:2 (1989), 293; R. Palmer, *Land and Racial Domination in Rhodesia* (London, 1977), p. 202.
8 See McGregor, 'Woodland resources', pp. 95-107; Drinkwater, 'The state and agrarian change', pp. 77-82.
9 Palmer, *Land*, p. 202; Steele, 'The foundations of a "Native" policy', p. 417.
10 W. Beinart, 'Soil erosion, conservationism and ideas about development: a southern African exploration, 1900-1960', *Journal of Southern African Studies*, 11:1 (1984), 75-7.
11 National Archives of Zimbabwe, Harare (hereafter NAZ), S1563, *Annual Report of the Director of Native Agriculture for the Year 1945*.
12 NAZ, S1563, NC's Annual Report for 1947, Wedza. The NC complains at length in his report of the demands of enforcing conservationist policies.
13 See K. Wilson, 'History, ecology and conservation in Southern Zimbabwe', Oxford, MS, 1986.
14 *Report of the Native Production and Trade Commission* [Godlonton Commission], Chairman W. A. Godlonton (Salisbury, 1944), p. 36.
15 Drinkwater, 'Technical development and peasant impoverishment', p. 300.
16 M. Yudelman, *Africans on the Land* (Cambridge, Mass., 1964), p. 117.
17 Quoted in I. Phimister, 'Rethinking the reserves: Southern Rhodesia's Land Husbandry Act reviewed', *Journal of Southern African Studies*, 19:2 (1993), 228.
18 Drinkwater, 'Technical development and peasant impoverishment', p. 285.
19 See the analyses of the interests of farming capital and secondary industry in W. Duggan, 'The Native Land Husbandry Act of 1951 and the rural African middle class of Southern Rhodesia', *African Affairs*, 79 (1980); G. Arrighi, *The Political Economy of Southern Rhodesia* (The Hague, 1967); Phimister, 'Rethinking the reserves'; and I. Phimister, *An Economic and Social History of Zimbabwe, 1890-1948* (London, 1986), pp. 270-2.
20 Holleman, *Chief, Council and Commissioner*, pp. 54-5, 31-2.
21 Drinkwater, 'The state and agrarian change', p. 103, and see pp. 67-8, 88-9.
22 Holleman, *Chief, Council and Commissioner*, pp. 38-9, 33. Holleman played a key role on the Mangwende Commission.
23 Quoted in Holleman, *Chief, Council and Commissioner*, p. 39, n. 2.
24 Godlonton Commission, p. 46.
25 NAZ, S1217/9, Native reserves land utilisation and good husbandry bill, Introductory note by A. Pendered (Marketing Officer), 10 April 1948, quoted in Phimister, 'Rethinking the reserves', p. 231. The sentiment that labour migration led to 'grossly inefficient' farmers and workers was reiterated repeatedly in the next decade. E.g. see NAZ, 6.5.9R/84273, F. H. Dodd, Administrative Officer, NLHA, 'The Native Land Husbandry Act', October 1958.

26 NAZ, 6.5.9R/84273, F. H. Dodd, Administrative Officer, NLHA, 'The Native Land Husbandry Act', October 1958. Also see *Annual Report of the Director of Native Agriculture for the Year 1958*, R. M. Davies, March 1959.
27 Quoted in the *Rhodesian Herald*, 19 May 1955.
28 *Annual Report of the Secretary for Native Affairs for the Year 1946*, p. 2. For a summary of the pressures of evictions and the great concern they caused administrators, see NAZ, 14.8.8F/69691, 'Native land position, summary of ad hoc reports', January 1947.
29 K. Wilson, 'Research on trees in the Mazhiwa and surrounding areas', ENDA-Zimbabwe, Harare, MS, 1987, p. 59.
30 Yudelman, *Africans on the Land*, p. 119.
31 See M. Chanock, 'Paradigms, property: A review of the customary law policies, and of land tenure', in K. Mann and R. Roberts (eds), *Law in Colonial Africa* (London, 1991), pp. 61-84, on the very widespread colonial desire to subordinate African 'rights' to land to the imperatives of 'development', however defined.
32 See Low and Lonsdale, 'Introduction', p. 13.
33 Drinkwater, 'The state and agrarian change', pp. 89-90. And see M. E. Bulman, *The Native Land Husbandry Act of Southern Rhodesia: A failure in land reform* (Salisbury, 1973), Appendix C and Appendix F; Yudelman, *Africans on the Land*, pp. 120-2.
34 The first use of aerial photography in classifying different types of land was in the Nata reserve in 1949-50. H. Weinmann, *Agricultural Research and Development in Southern Rhodesia, 1924-50* (Salisbury, 1975), p. 208.
35 NAZ, 23.7.5R/93142, Internal Affairs Correspondence Files, Per 5/GEN, L. Powys-Jones, CNC, File No. 542, Circular No. 322, 1 July 1951. The policy was partially reversed in 1957 due to the resentment it caused among chiefs.
36 NAZ, 23.7.5R/93142, Per 5/GEN, P. B. Fletcher, Ministry of Native Affairs, Memorandum by the Minister of Native Affairs, Proposed increases in Chiefs' and Headmen's subsidies and personal allowances, 21 October 1957.
37 *Ibid.*
38 NAZ, 23.7.5R/93142, Per 5/GEN, Memorandum by the Ministry of Native Affairs on increased subsidies for Chiefs and Headmen, Salisbury, n.d. [1957].
39 Bulman, *The Native Land Husbandry Act*, and others have stressed the conservationist motives behind the Act. Phimister, 'Remaking the reserves', pp. 227, 232, shows that the Cabinet's decision to speed the Act's implementation was motivated very directly by concern over the allegedly dire state of the reserves.
40 See, *inter alia*, B. N. Floyd, 'Changing patterns of African land use in Southern Rhodesia', Ph.D. dissertation, Syracuse University, 1959; Duggan, 'The Native Land Husbandry Act'; Bulman, *The Native Land Husbandry Act*; I. Scoones, 'Livestock populations and the household economy: A case study from Southern Zimbabwe', Ph.D. dissertation, University of London, 1990.
41 Bulman, *The Native Land Husbandry Act*, pp. 33, 12.
42 Drinkwater, 'The state and agrarian change', p. 92. Yields estimated by NCs and demonstrators were highly inaccurate. See Central Statistical Office, *Sample Survey of African Agriculture* (Salisbury, 1962).
43 *Second Report of the Select Committee on Resettlement of Natives* [Quinton Report], Salisbury, 1960, paras 130-7.
44 Drinkwater, 'The state and agrarian change', p. 93.
45 NAZ, 6.5.9F/84267, Native Land Husbandry Act, Standing Committee, Minutes of the Sixth Meeting, 7 January 1956. See also Bulman, *The Native Land Husbandry Act*, p. 13; NAZ, 6.5.9R/84273, F. H. Dodd, Administrative Officer, NLHA, 'The Native Land Husbandry Act', October 1958.
46 That the Act undermined security of tenure is widely accepted. See Bulman, *The Native Land Husbandry Act*, pp. 22-3; Holleman, *Chief, Council and Commissioner*, pp. 62-3; Floyd, 'Changing patterns of land use', pp. 119-20; *Report of the*

Mangwende Reserve Commission of Inquiry [Mangwende Commission], Chairman J. S. Brown, Salisbury, 1961, paras 82–90.
47 Floyd, 'Changing patterns of land use', ch. 8, shows that there were far too few stock to maintain soil fertility under the assumptions of the Alvord production system in a wide variety of reserves and, pp. 303–4, estimates that over sixty per cent of right holders would have less than the recommended number of stock, many having none. Officials were not unaware of these problems. See NAZ, 6.5.9R/84273, *Annual Report of the Director of Native Agriculture for the Year 1958*, R. M. Davies, Salisbury, March 1959, p. 5, for example.
48 *Ibid.*, p. 8. See also, e.g., NAZ, 6.5.6R/84266, PNC, Mashonaland East, to Under Secretary, Native Agriculture and Land, 28 January 1960.
49 NAZ, 6.5.9R/84273, *Annual Report of the Director of Native Agriculture for the Year 1958*, R. M. Davies, Salisbury, March 1959, pp. 4, 38–40.
50 J. Alexander, 'The state, agrarian policy and rural politics in Zimbabwe: Case studies of Insiza and Chimanimani District, 1940–1990', D.Phil. dissertation, Oxford University, 1993, pp. 33–5.
51 NAZ, 27.6.6F/100842, LAN/20/GEN, R. L. C. Cunliffe, 'Annual Report. Native Land Husbandry Act', Salisbury, May 1963. Cunliffe cites figures showing a sixteen per cent growth in population and a fifteen per cent growth in employment between 1951 and 1956 in contrast to an eighteen per cent growth in population and five per cent growth in employment between 1956 and 1961.
52 These problems have been discussed by, *inter alia*, Bulman, *The Native Land Husbandry Act*, pp. 16, 21; Holleman, *Chief, Council and Commissioner*, p. 63; Drinkwater, 'The state and agrarian change', p. 97; Yudelman, *Africans on the Land*, p. 131.
53 See NAZ, 6.5.9F/84267, Native Land Husbandry, Standing Committee, Minutes of the Eighth Meeting, 15 April 1957; Minutes of the Ninth Meeting, 3 June 1957; Minutes of Ad Hoc Meeting to discuss the common levy and destocking, 25 June 1957; R. L. C. Cunliffe, NLHA, Destocking, Memorandum prepared for the Native Land Husbandry Act Standing Committee, June 1957; R. M. Davies, Memorandum on Destocking in Native Areas presented to the Native Land Husbandry Standing Committee, 25 June 1957.
54 On irrigation schemes, see NAZ, 6.1.9F/84256, S. M. Makings, Department of Native Economics and Marketing, 'Manicaland irrigation schemes: Economic investigation', June 1958, and 6.5.6R/84266, Assistant Secretary to Administrative Officer, NLHA, 14 August 1956. On low rainfall areas, see 18.8.10R/88379, Arthur Hunt, Native Affairs Department Economist, 'Cattle in the lower rainfall regions of Southern Rhodesia', 26 February 1959, and Bulman, *The Native Land Husbandry Act*, p. 15. For higher rainfall areas, see questions raised by the Manicaland Provincial Agriculturalist and Arthur Pendered, Under Secretary, Native Economics and Marketing in 1957, cited in Bulman, *The Native Land Husbandry Act*, p. 20, n 1.
55 NAZ, 6.5.9R/84273, S. M. Makings, Department of Native Economics and Marketing, 'The problem of the communal grazings', Salisbury, 13 October 1958, p. 11.
56 Central Statistical Office, *Sample Survey*, and see Holleman's discussion in *Chief, Council and Commissioner*, pp. 325–8.
57 See N. Kriger, *Zimbabwe's Guerrilla War: Peasant voices* (Cambridge, 1992), pp. 110–15; R. Weitzer, 'In search of regime security: Zimbabwe since independence', *Journal of Modern African Studies*, 22:4 (1984); N. Bhebe, 'The nationalist struggle', in C. Banana (ed.), *Turmoil and Tenacity: Zimbabwe 1890–1990* (Harare, 1989), pp. 99–109; D. Martin and P. Johnson, *The Struggle for Zimbabwe: The Chimurenga war* (Harare, 1981), pp. 60–8.
58 The NAAB was established in 1941, and staffed by the Ministry of Native Affairs' senior officials. See Murray, *The Governmental System*, p. 312.
59 NAZ, 6.1.9F/84256, C. J. Bisset, Acting Under Secretary, Native Agriculture and Lands, 'Working party on the immediate future of the Land Husbandry Act', Annexure E to Special NAAB Meeting, Salisbury, 20–22 March 1961.
60 NAZ, 6.1.9F/84256, Special NAAB Meeting, Salisbury, 20–22 March 1961.

61 *Ibid.*
62 See Bhebe, 'The nationalist struggle', pp. 70–84, 98–108.
63 NAZ, 6.1.9F/84256, Special NAAB Meeting, Salisbury, 20–22 March 1961.
64 NAZ, 6.1.9F/84256, C. J. Bisset, Acting Under Secretary, Native Agriculture and Lands, 'Working party on the immediate future of the Land Husbandry Act', Annexure E to Special NAAB Meeting, Salisbury, 20–22 March 1961.
65 NAZ, 6.1.9F/84256, 'The Native Land Husbandry Act and present problems of political agitation', Annexure B to Special NAAB Meeting, Salisbury, 20–22 March 1961.
66 Paragraph based on NAZ, 6.1.9F/84256, Arthur Pendered, Deputy Secretary, Development, 'departmental organization for African agricultural development and implementation of the Native Land Husbandry Policy', Salisbury, 25 January 1961, Annexure A to Special NAAB Meeting, Salisbury, 20–22 March 1961. For a clear statement of the administrative perspective, see NAZ, 27.6.6F/100842, PNC Campbell, Bulawayo, to Secretary for Internal Affairs S. E. Morris, 7 September 1960.
67 See the Quinton Report, and Gloria Passmore, *The National Policy of Community Development in Rhodesia* (Salisbury, 1972), pp. 71–4.
68 Passmore, *The National Policy of Community Development*, p. 74. See Holleman, *Chief, Council and Commissioner*, p. 331, n. 1, on Green's extensive influence. The policy of community development originated in a 1948 conference of colonial administrators at Cambridge University; in 1963, forty-seven countries were implementing community development programmes with USAID backing. See NAZ, 6.7.5F/84279, James Green, *Community Development in Southern Rhodesia*, 1963.
69 See Holleman, *Chief, Council and Commissioner*, pp. 255–91; Passmore, *The National Policy of Community Development*, pp. 162–84.
70 See NAZ, 27.6.6F/100842, R. L. C. Cunliffe, Annual Report, Native Land Husbandry Act, May 1963; 6.1.9F/84256, Special NAAB Meeting, Salisbury, 20–22 March 1961; 6.7.5F/84279, Roger Howman, Deputy Secretary, Internal Affairs, 'The application of the philosophy of community development in Southern Rhodesia', 1963; and Passmore, *The National Policy of Community Development*, pp. 48–50, on the 'opposing claims of technical and human factors'. Green played a key role in introducing the 'human factor' discourse into the reports of the Mangwende and Robinson Commissions.
71 *Report of the Secretary for Native Affairs and Chief Native Commissioner for the Year 1961*, Salisbury, 1962, pp. 26–7.
72 See NAZ, 6.1.9F/84256, C. J. K. Latham, District Officer/Delineation Officer, 'Community development: Methods and techniques', July 1963; 'Delimitations of communities', Senga Community Development Centre, Gwelo, 1963; 28.10.8F/98431, J. W. Green, 'Community development in Southern Rhodesia', Paper delivered to the Southern Rhodesia Information Service, March 1963; 6.1.9F/84256, James W. Green, Community Development Adviser, 'Prerequisites to African development – The Three R's', 1 June 1961. See also Passmore, *The National Policy of Community Development*, p. 95.
73 Four Working Parties were established under the Robinson Commission. Working Parties A, B and C made recommendations regarding non-racial administration, a new department of African agriculture and tribal courts. All were extremely influential in shaping policy change.
74 The Working Party's membership included Under Secretary, Administration, Roger Howman, an advocate of community development, Rhodesia's leading authority on local government and African courts, and the architect of the African Councils Act of 1957, as well as later legislation on courts; R. Cunliffe, Administrative Officer for the NLHA; two prominent NCs; and community development adviser James Green.
75 NAZ, 6.1.9F/84256, Minutes of the Meeting of the NAAB, Salisbury, 27–28 June 1961.
76 NAZ, 14.8.8F/69691, Working Party D, Arising from the Robinson Commission

Report, 'The tribal authority and the land', Paper No. 16, Salisbury, 2 September 1961.
77 NAZ, 14.8.8F/69691, Working Party D, 'Revision of the Native Land Husbandry Act', Paper No. 23, 24 December 1961.
78 See NAZ, 14.8.8F/69691, Working Party D, 'The tribal authority and the land', Paper No. 16, Salisbury, 2 September 1961.
79 See Quinton Report, paras 346–50, 232–54.
80 NAZ, 14.8.8F/69691, Working Party D, 'The tribal authority and the land', Paper No. 16, Salisbury, 2 September 1961.
81 It is beyond the scope of this chapter to explore fully the critiques which Africans made of the NLHA. Suffice it to say that those who sat on Assessment Committees, or who offered objections at meetings, through the nationalist parties or elsewhere, were often aware of the flaws which bedevilled the agricultural science on which the Act was based. There was also a sophisticated assessment of which sorts of objections were most likely to be effective in the face of official expertise: Africans were willing to play the 'spiritual' card where they deemed it advantageous. Needless to say, awareness of the way in which land redistribution and national politics were kept neatly off the agenda was a constant source of resistance and resentment.
82 *Report of the Advisory Committee on the Economic Development of Southern Rhodesia with particular reference to the role of African agriculture* [Phillips Commission], Chairman John Phillips, Salisbury, 1962, pp. 148–52. Emphasis in original.
83 NAZ, 14.8.8F/69691, Working Party D, 'The tribal authority and the land', Paper No. 16, Salisbury, 2 September 1961.
84 NAZ, 14.8.8F/69691, Working Party D, 'Revision of the Native Land Husbandry Act', Paper No. 23, 24 December 1961.
85 *Ibid.*
86 *Ibid.*
87 The Working Party defined the Large Stock Equivalent as including all bovines, horses, donkeys and mules over twelve months of age. NAZ, 5.2.8R/82725, Working Party D, Third Report, District Surveys of Land, Population and Stock, Introductory Chapter, 26 September 1962. According to J. D. Jordan, 'Zimutu Reserve: A land use appreciation', *Rhodes Livingstone Institute Journal*, 36 (1964), 69, n. 1, the Annual Cropping Acreage concept was first proposed by Conservation Planning Officer P. A. Colville in 1961.
88 *Ibid.* Working Party D calculated on the basis of 'U' factors – land taken up by conservation works, roads and rocks – and 'F' factors, the minimal rotation which would allow the maintenance of soil fertility. The calculation of the 'F' factor is unclear. No farming system is specified, nor is the use of natural or artificial fertilisers. Annual Cropping Acreages calculated by Jordan, 'Zimutu Reserve', pp. 69–70, varied widely depending on the system used to maintain fertility.
89 NAZ, 5.2.8R/82725, Working Party D, District Survey, Makoni, October 1962.
90 NAZ, 5.2.8R/82725, Working Party D, Third Report, District Surveys of Land, Population and Stock, Introductory Chapter, September 1962, and, e.g., Working Party D, District Survey, Insiza (Filabusi), 11 July 1963.
91 NAZ, 14.8.8F/69691, Working Party D, 'Revision of the Native Land Husbandry Act', Paper No. 23, 24 December 1961; 25.10.6R/100839, Deputy Secretary, Development, W. H. H. Nicolle to Provincial Commissioner, Matabeleland North, 19 December 1961; 6.1.9F/84256, Draft Report of Working Party B, Establishment of Department of Agriculture, Working Paper on 'Land husbandry planning and implementation', 14 September 1961.
92 R. Howman, 'Chieftainship', *NADA*, 1966, quoted in T. Ranger, 'Democracy and traditional political structures, 1890–1995', Oxford, MS, p. 21. Howman was of course one of the key architects of local government policy. Unlike many of his Rhodesian Front colleagues, he cautioned against the dangers of an overly zealous search for tradition.

93 Secretary for Internal Affairs Circular, February 1973, cited in T. Ranger, 'Zimbabwe and the long search for independence', in D. Birmingham and P. Martin (eds), *History of Central Africa: The contemporary years since 1960* (London, 1998), p. 208.
94 See Ranger, 'Zimbabwe', pp. 207–8 and 'Democracy', pp. 20–2; Alexander, 'The state, agrarian policy and rural politics', ch. 3.

INDEX

Note: 'n' after a page reference indicates a note number on that page.

Adas, Michael 101
administrative centralisation 100–1, 102, 118–20, 123, 125, 159, 213–15, 224–5
  see also technocrats
African nationalism 4, 84, 191, 215, 219, 222–3, 226, 236n.81
Afrikaner Bond 104
Afrikaner nationalism 82, 88–9, 90–2, 94, 125–6, 206–7
Agassiz, Louis 12, 13, 27, 28
agriculture 42–63, 100–12, 212–31 passim
  in African reserves 212, 213–14, 215–31 passim
  science of 42–63 passim, 77, 106–7, 214–31 passim
  see also peasants
Alvord, E. D. 214, 215, 217
anthropology 2, 4, 11–12, 27, 30–1, 73, 81, 87, 157–8, 179, 228
  cultural relativism 30–1, 179, 193, 197
apartheid 5, 116–17, 125–7, 128–38 passim, 201, 206–7
  see also segregation
Ashforth, Adam 121
astronomy 10n.10, 67, 73

Barber, Mary 26–7
Barbey, William 17, 19, 36n.19
Baumann, Zygmunt 118, 119
Bews, J. W. 194
birth control 6–7, 164–83 passim
Boas, Franz 16, 30
Bolus, Harry 78
Bonâme, Phillipe 48, 53, 54–5, 56–7, 58

botany 14, 18–21, 33–4, 78–9, 81
British Association for the Advancement of Science 5, 66–94
Britten, Henry 169, 175
Brockway, Lucile 63
Bryce, James (Lord) 25

Cape Colony 4, 100–12
Cape Town Mother's Clinic 177
Carnegie Commission (1932) 88, 121, 170, 172
Carr, W. J. P. 137
Caton Thompson, Gertrude 87–8
census enumeration 105, 120–1, 123, 130, 132–4, 172, 218
Clarke, Fred 88
Cluver, E. H. 180, 189, 195, 196, 198, 199
collecters 18–23
colonial state 44, 45, 54, 63, 70, 108, 100
commonwealth 5, 6, 66, 83, 86
conservation 214, 217, 218, 220, 233n.39
Council for Scientific and Industrial Research 124
Currie, Donald 67

Dart, Raymond 81, 87, 88
Darwin, George 66, 72–3
de Villiers, I. P. 150–1, 152, 153
dogs 5, 143–59
Drinkwater, Michael 215
Dubow, Saul 43, 138
Durban Medical School 20, 35, 193, 197, 203–5
  see also medicine

[238]

INDEX

Eiselen, W. M. 129
entomology 17, 21, 22–3, 25–6, 33–4
eugenics 2, 6, 164–83 *passim*
 see also scientific racism
evolution 12–15, 25, 26–7, 31, 81, 91, 94

Fagan, H. A. 121–2, 124–5
Fantham, H. B. 166–8, 171, 173, 180, 181, 182
Foucault, Michel 2, 117, 199
Frobenius, Leo 87

Gale, George 6, 189–95, 196–207 *passim*
gender 131, 133, 141n.70
 see also women
geodesy 73
geology 13, 16–18, 73, 81
Geyer, Albert 92
Gill, David (Sir) 67, 69–70, 76, 81
Gluckman, Henry 180, 188, 189
 and National Health Services Commission 188, 190, 193, 199–203
Graham, Thomas (Sir) 147, 148, 155
Green, James 224, 225, 226, 228

Haddon, A. C. 73
Hamilton, Carolyn 43
Hardy, J. L. 169
Hely-Hutchinson, Walter (Sir) 70
Herschel, John (Sir) 95n.17
Hertzog, J. B. M. 83, 91, 93–4
Hoërnle, R. F. A. 176, 177–8
Hoërnle, Winifred 176, 177
Hoffman, Josias 105
Hofmeyr, Jan 6, 84, 85, 194, 203
Holland, Thomas (Sir) 83, 86
Holleman, J. F. 216
Honwana, Raúl 33
Horne, John 45, 46, 48–9, 51
Hunt, Arthur 221
Hunter, Monica 157–8

Icery, Edmond 47
imperialism 66–7, 69, 85, 92, 93, 101
institutions 12–13, 21, 22–3, 47–8, 49–50, 62, 74–80, 121
 see also individual entries
intelligence quotient (IQ) 177–8

Jebb, Richard 71, 74
Jock of the Bushveld 83
Junod, Henri-Alexandre 3, 11–35 *passim*

Kark, Emily 188, 189, 190, 195–6
Kark, Sydney 6, 188, 189, 190, 193, 195–6, 198, 200, 204, 205–6
Kew Gardens 45, 47, 48, 53, 63, 78, 79
Kirstenbosch Botanical Gardens 77, 78–9, 97n.60
knowledge 2, 3, 5, 10n.7, 46, 70, 90, 102
 applied 2, 47–8, 60, 72, 75, 86, 89
 imperial 3, 7, 43, 66
 indigenous systems 7, 23, 30, 110–12, 157–8
 magic and superstition 8, 33, 104, 157, 193, 196–7
 as power 2–3, 6, 9, 62–3, 116, 118–19
 scientific 2, 5, 7–8, 9, 33, 62–3, 68, 230

labour 44, 122, 128–9, 131–2, 136, 168, 178, 207
migrancy 192, 195, 221
Leclézio, Henri 50, 53, 58
Libombo, Elias 23–4, 30

Makings, S. M. 221
Malan, D. F. 90, 170
Malherbe, E. G. 88, 89, 90, 121, 122, 203–4
Mamdani, Mahmood 125–6
Mansfield, Edward 168, 182
Mauritius 8–9, 42–63 *passim*

[239]

INDEX

Medical Association of South Africa 200
medicine 2, 74, 77, 169–70, 180, 188, 202
  social medicine 188–207 *passim*
  see also Durban Medical School
Merriman, John X. 103–4
Milner, Alfred (Lord) 67, 76–7, 93
modernity 5, 76–7, 101–2, 116–17, 124, 137–8, 157, 159, 218
  resistance to 104, 110–11, 230–1
  see also progress
Moll, A. M. 170
Moroka, James 191
Morris, S. E. 224, 225
Mozambique 16, 18–33 *passim*
museums 17, 22, 23, 75, 77–8

National Health Services Commission *see* Gluckman, Henry
nationalism 5, 9, 74–5, 80
  in settler colonies 5, 76, 78, 84, 93–4, 151
'native administration' 116, 120, 124, 127–8, 129, 132, 135–7, 216–18
  technical development 212–31 *passim*
Native Land Husbandry Act (1951) 4, 213–14, 215–22, 223–31 *passim*
Natives (Urban Areas) Act (1923 and 1952) 122–3, 128, 178
Naz, Virgile 46–7, 58
Neuchâtel 12–18, 34
Newton, William 48, 49

Onderstepoort 77

Pamplemousses Botanic Gardens 45, 47, 50, 51, 52, 53, 55, 57, 62
Pearson, Harold 78
peasants 42, 52, 60, 61, 212, 214, 215, 218, 220–1, 222–31 *passim*
  see also agriculture

Pendered, Arthur 216, 217, 223–4, 229
Péringuey, Louis 22, 73
Perromat, Georges 50–1, 55, 56–7
Pholela Health Centre 188, 190, 195–6, 200
'poor whiteism' 6, 88, 89, 164, 170, 172–3, 174, 181, 202
Population Registration Act (1950) 128
Porter, Annie 166–8, 172, 180, 181, 182
Porter, Theodore 102
poverty 195, 168, 172–3, 198–9
progress 93, 100–12
  see also modernity

race discrimination 60, 83, 94, 126, 128, 149, 151, 176–7, 178–9, 182–3, 231
Race Welfare Society 6–7, 164–83
Radcliffe-Brown, A. R. 34–5, 179
Randall-McIver, David 73
religion 12, 13–16, 18, 169, 170
Reunert, Theodore 69, 75
Rhodesia (Southern) 72, 73, 212–31
Rockefeller Foundation 188, 190, 193, 203, 204, 205
Rothmann, M. E. 170
Rutherford, Ernest (Sir) 82, 83
Ryle, John 189, 202

Sadie, J. L. 127–8
Scab Act and Commission 104, 110, 111
scientific racism 32, 81–2, 90, 149
  see also eugenics
Schaffer, Simon 46, 61–2
Schlechter, Rudolf 20
Seeley, John 101
segregation 34, 82, 90, 197–8, 213
  see also apartheid
Shapin, Steven 46, 61–2
Smuts, Jan 6, 80–1, 84, 85–6, 87, 189, 194, 201, 202
social welfare 169, 177, 180, 206

INDEX

Social and Economic Planning Council 124
South African Association for the Advancement of Science 5, 66–94 *passim*
South African Institute for Medical Research 167
South African Institute of Race Relations 178, 179
South Africanism 5, 78–9, 84, 88, 151
South African National Council of Women 175
South African Philosophical Society (Royal Society) 74, 96n.42
South African Police 144, 145, 148, 150–2, 153–4
statistics 4, 59, 100–12 *passim*, 116–38 *passim*, 221–2
sugar cane 8–9, 42–63

technocrats 4, 77
  development policies 212–15, 216–22, 224–6, 321–2
  and 'human factor' 222–5
  see also administrative centralisation

technology 9, 76, 93
Theiler, Arnold 77
Theron, Thomas 104
Thorndike, E. L. 122
Thornton, Edward 174, 175
tribalism 27, 28–30, 131, 213, 216–17
  chieftancy 219, 226–7
  custom 217–18, 230–1
Tripet, Fritz 14
Truter, T. G. 150, 158

universities 17, 79–80, 84, 194, 195, 197, 203–5
urbanisation 122–3, 124, 128, 168, 213

Verwoerd, Hendrik 127, 137

Weber, Max 118
women 24, 166, 170, 171, 172, 174, 175, 176–9
  see also gender
Woodrow, Elsa 176, 177
Worboys, Michael 68–9

Xuma, A. B. 191

EU authorised representative for GPSR:
Easy Access System Europe, Mustamäe tee 50,
10621 Tallinn, Estonia
gpsr.requests@easproject.com